REVISE EDEXCEL GCSE (9–1)
Geography B

Series Consultant: Harry Smith

Author: Andrea Wood

A note from the publisher

In order to ensure that this resource offers high-quality support for the associated Pearson qualification, it has been through a review process by the awarding body. This process confirms that this resource fully covers the teaching and learning content of the specification or part of a specification at which it is aimed. It also confirms that it demonstrates an appropriate balance between the development of subject skills, knowledge and understanding, in addition to preparation for assessment.

Endorsement does not cover any guidance on assessment activities or processes (e.g. practice questions or advice on how to answer assessment questions), included in the resource nor does it prescribe any particular approach to the teaching or delivery of a related course.

While the publishers have made every attempt to ensure that advice on the qualification and its assessment is accurate, the official specification and associated assessment guidance materials are the only authoritative source of information and should always be referred to for definitive guidance.

Pearson examiners have not contributed to any sections in this resource relevant to examination papers for which they have responsibility.

Examiners will not use endorsed resources as a source of material for any assessment set by Pearson.

Endorsement of a resource does not mean that the resource is required to achieve this Pearson qualification, nor does it mean that it is the only suitable material available to support the qualification, and any resource lists produced by the awarding body shall include this and other appropriate resources.

> For the full range of Pearson revision titles across KS2, KS3, GCSE, Functional Skills, BTEC and AS/A Level visit:
> www.pearsonschools.co.uk/revise

Contents

• • • • • • • • • • • • • • • • • • • •

Edexcel publishes Sample Assessment Material and the Specification on its website. This is the official content and this book should be used in conjunction with it. The questions have been written to help you practise every topic in the book. Remember: the real exam questions may not look like this.

1-to-1 page match with the Geography B Revision Guide ISBN 9781292133782

Global circulation

1   Analyse the data in **Figure 1**.

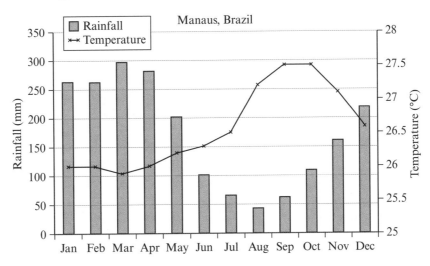

Figure 1   A climate graph for Manaus, Brazil, located 3° south of the equator

Which of the following describes Manaus' climate?

   ☐   **A**   Hot dry arid   ☐   **C**   Cool wet winters, hot dry summers

   ☐   **B**   Hot wet tropical   ☐   **D**   Warm wet summers, cool dry winters   **(1 mark)**

2   Calculate the temperature range for Manaus.

> The temperature range is the difference between the highest and lowest temperatures.

....

....   **(1 mark)**

3   Analyse the data in **Figure 2**.

Location	Latitude	Days snow	Average day temp	Average night temp
Tresco, Isles of Scilly, UK	49.5°N	0	10°C	4°C
Newfoundland, Canada	47.5°N	18	−5°C	−9°C

Figure 2   Climate information for two places in the mid latitudes

Calculate the difference between the average daytime temperatures.

....   **(1 mark)**

> **Guided**

4   Compare the climates of Tresco and Newfoundland.

> To **compare** you need to look for similarities and differences between the two items. Try to use comparative vocabulary, such as 'whereas'.

Tresco and Newfoundland are located at similar

latitudes but their climates are very different.

Tresco is much ....

....

....

....   **(3 marks)**

1

Natural climate change

1 Study **Figure 1**.

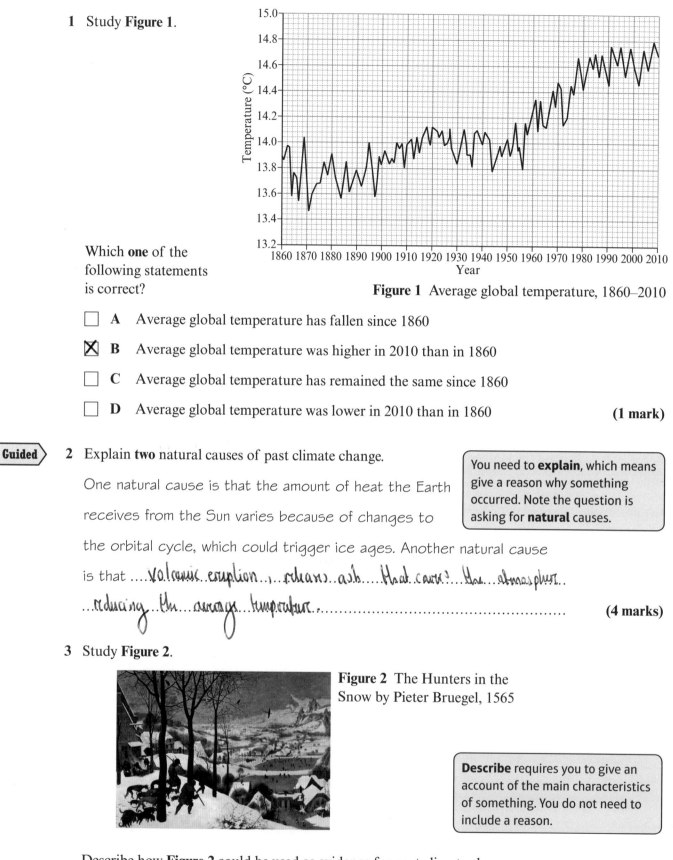

Figure 1 Average global temperature, 1860–2010

Which **one** of the following statements is correct?

☐ A   Average global temperature has fallen since 1860

☒ B   Average global temperature was higher in 2010 than in 1860

☐ C   Average global temperature has remained the same since 1860

☐ D   Average global temperature was lower in 2010 than in 1860

(1 mark)

> Guided

2 Explain **two** natural causes of past climate change.

> You need to **explain**, which means give a reason why something occurred. Note the question is asking for **natural** causes.

One natural cause is that the amount of heat the Earth receives from the Sun varies because of changes to the orbital cycle, which could trigger ice ages. Another natural cause is that ....volcanic eruption.., releans ash....that cause? the atmosphere. ..reducing the average temperature..................................... **(4 marks)**

3 Study **Figure 2**.

Figure 2 The Hunters in the Snow by Pieter Bruegel, 1565

> **Describe** requires you to give an account of the main characteristics of something. You do not need to include a reason.

Describe how **Figure 2** could be used as evidence for past climate change.

....there is snow suggesting it was a cold climate..........
....
.... **(2 marks)**

Humans and climate change

1 Burning fossil fuels in power stations is thought to contribute to the **enhanced greenhouse effect**. Explain **two** other types of human activity that are thought to contribute to the enhanced greenhouse effect.

... Agriculture, produces methane via cows ....

....

- Deforestation, distruction of tres causing less oxygen production

for demstic production.   **(4 marks)**

Guided 2 Study the following extract from a report on climate change in Bangladesh. It shows some of the predicted range of impacts from climate change on Bangladesh by 2100.

> Climate change can affect food production. Crop yields depend on temperature and rainfall and they may be reduced by up to 30 per cent. Cyclones will become stronger, with faster winds causing more damage; floods will become more common and, with a rise in sea level, many low-lying areas will be flooded. This will lead to increased amounts of disease among the population.

Describe **one** piece of evidence for recent climate change.

Recent extreme weather events, such as ...... a cyclone incrased

..... cyclones have incrased strngth & speed   **(2 marks)**

3 Study **Figure 1**.

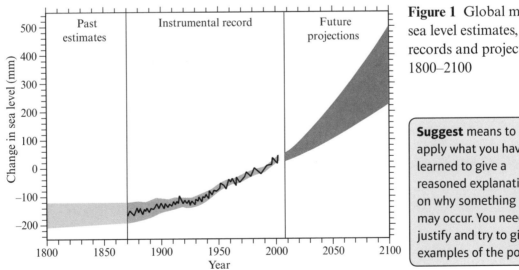

Figure 1 Global mean sea level estimates, records and projections, 1800–2100

> **Suggest** means to apply what you have learned to give a reasoned explanation on why something may occur. You need to justify and try to give examples of the point.

Suggest **one** reason why there is such a large difference between the projected maximum and minimum sea levels in 2100.

... Because we don't know what might change the max is i.f nothing

.... changes the minmum would be If we charge how

we live (becaus more sustainable)   **(2 marks)**

Tropical cyclones

1 Study **Figure 1**.

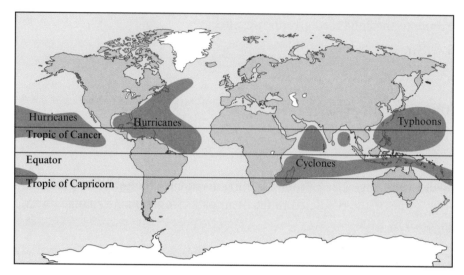

Figure 1 World distribution of tropical cyclones

Describe the distribution of tropical cyclones shown in the map.

> Try to use the equator as a way of describing the distribution.

...Most of the cyclones lay along / in between the tropics of...........

...Cancer and Capricorn. They are all near the equator...........

.... **(2 marks)**

> **Guided**

2 Explain how tropical cyclones form. Refer back to **Figure 1**.

The map shows the tropical cyclones being largely restricted to the

tropics. There are three main reasons for this location. First, these

storms are powered by warm ocean temperatures – the seawater

needs to be above 26.5 °C and these temperatures are only

found in late summer and autumn in the tropics. Second, ...

....

....

.... **(3 marks)**

3 Which of the following places is most likely to be affected by tropical cyclones?

☐ **A** UK

☐ **B** Argentina

☒ **C** Australia

☐ **D** Russia

(1 mark)

Tropical cyclone intensity

1 Which of the following is used to measure the magnitude of a tropical cyclone?

   ☐ **A** The Richter scale

   ☐ **B** The Beaufort scale

   ☒ **C** The Saffir–Simpson scale

   ☐ **D** The Mercalli scale   **(1 mark)**

Guided

2 Analyse the data in **Figure 1**.

Tropical cyclone	Location	Highest wind speed	Height of storm surge	Pressure (mb)	Fatalities
Typhoon Haiyan November 2013	Philippines	315 km/h	4.23 m	895	6340
Hurricane Sandy October 2012	USA	185 km/h	5.20 m	946	285

Figure 1 Climate data from two recent cyclones

Using the information from **Figure 1**, suggest the likely damage caused by Typhoon Haiyan.

Using the scale, Typhoon Haiyan was classified as a category 5 cyclone, where the damage would be ...Destruction of homes / buildings. Destruction of environment from floods and wind pressure..........

....

.... **(2 marks)**

3 Explain **one** reason why some tropical storms intensify into cyclones.

> The temperature of the oceans is an important factor.

...The heat of the ocean causes the hot air to.......
...rise...
.... **(2 marks)**

4 Describe the differences between the **two** cyclones shown in Figure 1.

...The USA hurricane was much lower slower but it had a higher...
...storm surge than the phillipinos one..................................
.... **(2 marks)**

Tropical cyclone hazards and impacts

Guided ▷ 1 Explain the **physical** hazards associated with tropical cyclones.

> The question asks for the physical hazards. Just describe the natural weather conditions associated with tropical cyclones.

Tropical cyclones can bring extreme weather conditions such as changes in atmospheric pressure, ...

....

....

....

.... **(3 marks)**

Guided ▷ 2 Explain **one** reason why tropical cyclones present a particular danger to low-lying coastlines.

> You need to give one reason and develop this in your explanation.

Low-lying coastlines are vulnerable to ....

caused by ....

....

....

.... **(3 marks)**

3 Describe the human impact cyclones can have.

....

....

....

.... **(2 marks)**

4 Suggest why some countries are more vulnerable than others to the impact of tropical cyclones.

....

....

....

.... **(2 marks)**

Dealing with tropical cyclones

Study **Figure 1**.

Figure 1 New floodgate over Bayou, New Orleans

1   Which of the following describes what **Figure 1** shows, and how it might protect New Orleans from future cyclones?

☐ **A**   A dam that will collect flood water

☐ **B**   A barrier that will keep back storm surges

☐ **C**   A weather forecasting centre to predict the track of cyclones

☐ **D**   A rescue centre for people to evacuate to   **(1 mark)**

2   Which of the following is used to help forecast and track cyclones?

☐ **A**   Satellites

☐ **B**   A weather balloon

☐ **C**   Tiltmeter

☐ **D**   Flow meter   **(1 mark)**

3   Referring to a named example, describe **one** method used to prepare for future cyclones.

> Make sure you refer to a specific place and a specific method.

....

....

.... **(2 marks)**

⟩ **Guided** ⟩   4   Explain why the impact of tropical cyclones is greater in some places.

> You need to refer to how places of different levels of development are affected by tropical cyclones.

Developed countries have more money and resources to provide ....

....

....

.... **(3 marks)**

Located example Tropical cyclones

1   Referring to a named example, evaluate the effectiveness of methods used to prepare for and respond to future tropical storms.

> To answer this question:
> *   name a specific place and protection measure for that place
> *   you could refer to weather forecasting, satellite technology, warning and evacuation, and storm surge defences
> *   talk about the strengths and weaknesses, suggest alternatives and use any relevant data
> *   make a judgement on whether the measures have been successful
> *   summarise your ideas in a short conclusion.

....

....

....

....

....

....

....

....

....

....

....

....

....

....

....

....

....

....

....

....   **(8 marks)**

> In your exam you should write approximately one side of extended writing for an 8-mark question. Finish off your answer on your own paper.

Tectonics

1   Study **Figure 1**.

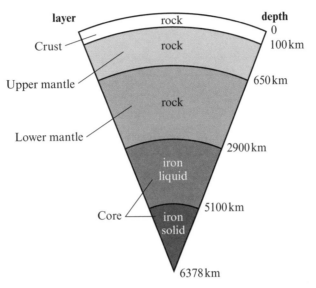

Figure 1 A cross-section showing the structure of the Earth

Which of the following is the thickest layer in the cross-section of the Earth?

☐ **A**   Crust

☐ **B**   Upper mantle

☒ **C**   Lower mantle

☐ **D**   Core

(1 mark)

> Guided

2   Compare the physical properties of the core and the mantle.

The inner core is solid, whereas ....the...outer...core...is.....liquid,..the...mantle..

...is.....also....solid..   **(2 marks)**

The inner core is hotter – over ......5000...degrees.....................................

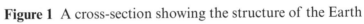

....   **(1 mark)**

3   Explain how convection currents contribute to plate movement.

| Circular motion is important to the movement. |

....

....

....

....

....

....   **(3 marks)**

Plate boundaries and hotspots

1   Study **Figure 1**.

Figure 1 Map showing the Earth's tectonic plates

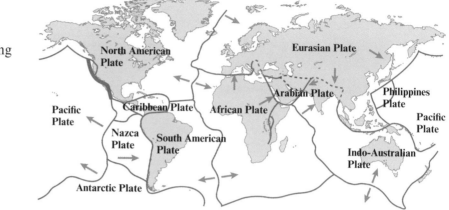

Which of the following is an example of a divergent plate boundary?

☒ **A**   Where the Eurasian Plate meets the North American Plate

☐ **B**   Where the Eurasian Plate meets the Pacific Plate

☐ **C**   Where the Pacific Plate meets the North American Plate

☐ **D**   Where the Nazca Plate meets the South American Plate   **(1 mark)**

2   Study **Figure 2**.

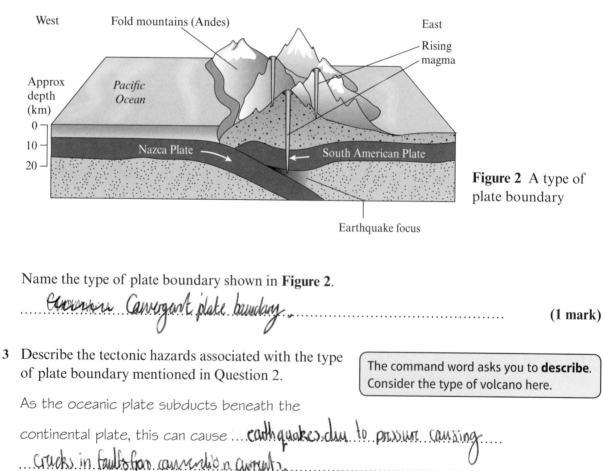

Figure 2 A type of plate boundary

Name the type of plate boundary shown in **Figure 2**.

...~~Corverno~~ Convergent plate boundary.. **(1 mark)**

> **Guided**

3   Describe the tectonic hazards associated with the type of plate boundary mentioned in Question 2.

> The command word asks you to **describe**. Consider the type of volcano here.

As the oceanic plate subducts beneath the

continental plate, this can cause ...earthquakes due to pressure causing....

...cracks in faults from convection currents..................................

....

.... **(3 marks)**

10

Tectonic hazards

1   Study **Figure 1**.

Figure 1
Mount Mayon,
the Philippines

Which of the following identifies the type of tectonic hazard shown in **Figure 1**?

☒ **A**   Composite cone volcano

☐ **B**   Shield volcano

☐ **C**   Rift valley

☐ **D**   Landslide   (1 mark)

2   Explain the tectonic processes that have created the landform shown in **Figure 1**.

> You can always use a diagram to illustrate the point you are trying to make.

A Convergence of tectonic plates ....

....

....

....   (3 marks)

Guided ⟩

3   Suggest **one** reason why the area shown in **Figure 1** may also be affected by earthquakes and other tectonic hazards.

When there is subduction of the oceanic plate under the continental

plate, this causes ....

....

....

....   (3 marks)

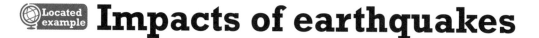

Had a go ☐ Nearly there ☐ Nailed it! ☐

Located example Impacts of earthquakes

Don't work through this page if you studied volcanoes for your located examples for this topic!

1 Study **Figure 1**.

1000s feared dead in Japanese quake

Yesterday afternoon an earthquake measuring 9.0 on the Richter scale hit Japan. The earthquake was followed by a powerful tsunami. Ten metre high waves washed inland destroying entire towns.

Rescue services are struggling to cope, with access to the worst hit areas proving almost impossible. Fires are burning in several cities and a number of nuclear reactors have been damaged, causing power shortages.

Figure 1 An article about the Japanese earthquake of 2011

Describe **one** secondary impact of the earthquake outlined in **Figure 1**.

....

.... **(2 marks)**

Secondary impacts are those which occur **after** the earthquake.

2 Which of the following is a primary impact of an earthquake?

☐ **A** Fires breaking out due to burst water pipes

☐ **B** Disease spreading as a result of contaminated water

☐ **C** Buildings collapsing

☐ **D** Power shortages

(1 mark)

Guided

3 Analyse the data in **Figure 2**.

Earthquake name and date	Location	Magnitude	GDP per capita (US$, 2012)	Number of deaths	Economic cost (US$)
Kathmandu (April 2015)	Nepal	7.8	2200	8700	10 billion
Tohoku (March 2011)	Japan	9.0	37 000	15 891	187 billion

Figure 2 The impact of two earthquakes

(i) Calculate the difference between the economic costs of the Kathmandu and Tohoku earthquakes.

.... **(1 mark)**

(ii) Explain **two** ways some earthquakes cause more damage and loss of life than others.

Earthquakes in some parts of the world often have a bigger impact than others because in poorer countries ....

....

Also the location affects the impact because ....

.... **(2 marks)**

🌐 Located example **Impacts of volcanoes**

> Don't work through this page if you studied earthquakes for your located examples for this topic!

Guided

1   Study **Figure 1**.

Figure 1 The Eyjafjallajökull volcano pumps ash into the sky, April 2010

State **two** secondary impacts often caused by volcanic eruptions.

One impact is that ash clouds from volcanic eruptions can stop aircraft

from flying, disrupting transport. Another impact is that people cannot

....

.... **(2 marks)**

2   Which of the following is a primary effect of a volcanic eruption?

☐ **A**   Climate cools as volcanic ash and dust are emitted into the atmosphere

☐ **B**   Roads are destroyed, blocked by lava flows

☐ **C**   Food shortages as a result of crops being destroyed

☐ **D**   Soils become more fertile from volcanic ash deposits   **(1 mark)**

3   Assess this statement, referring to named examples. 'Some places suffer more from the impacts of tectonic processes than others.'

....

....

....

....

....

....

.... **(8 marks)**

> For an **assess** question, you need to weigh up all the evidence and then make a conclusion about whether you agree with the statement. You need to back this up with evidence.

> In your exam you should write approximately one side of extended writing for an 8-mark question. Finish off your answer on your own paper.

🌐 Located example **Managing earthquake hazards**

> Don't work through this page if you studied volcanoes for your located examples for this topic!

1 Define the term 'short-term relief'.

....

....

....

           (1 mark)

2 Referring to named examples, state **two** ways people can protect themselves against earthquakes and volcanic eruptions.

....

....

....

....

....

           (3 marks)

3 For a named earthquake or volcanic eruption, explain the short-term relief efforts.

....

....

....

....

....

           (3 marks)

⟩Guided⟩   **4** One way to reduce damage from an earthquake is to design buildings in certain ways. With reference to a named example, explain how it can help minimise damage in an earthquake.

> Your answer must refer to **how** the technique reduces damage.

One example is putting automatic shutters on windows. These can

prevent ....

....

....

....

           (3 marks)

🌐 Located example Managing volcano hazards

Don't work through this page if you studied earthquakes for your located examples for this topic!

1   Study **Figure 1**.

Figure 1 Volcano evacuation route sign, near Mount Rainier, Washington, USA

Which of the following is this an example of?

☐   **A**   A short-term protection measure to provide emergency shelter for people affected by volcanic eruptions

☐   **B**   A tourist sign to encourage people to visit and learn about volcanoes

☐   **C**   A long-term planning measure to prepare for future eruptions

☐   **D**   A measure to keep people permanently away from the volcano   **(1 mark)**

⟩ **Guided** ⟩   2   Explain **one** short-term relief measure that can help people during volcanic eruptions.

During a volcanic eruption, rescue and evacuation centres can be

opened up away from the hazard, providing ....

....

....   **(3 marks)**

3   Referring to named examples, explain why some places are better prepared for volcanic hazards than others.

> Make sure you refer to specific places you have studied. You also need to make a comparison between places of different levels of development.

....

....

....

....

....   **(4 marks)**

Had a go ☐   Nearly there ☐   Nailed it! ☒

What is development?

1   Which of the following is a definition of GDP per capita?

☐ **A**   The total value of all the money in the banks of a country, divided by the population

☐ **B**   The total value of all the natural resources in a country, divided by the population

✓ ☒ **C**   The total value of goods and services produced by a country in a year, divided by the population

☐ **D**   The average income in the country

(1 mark)

2   Study **Figure 1**.

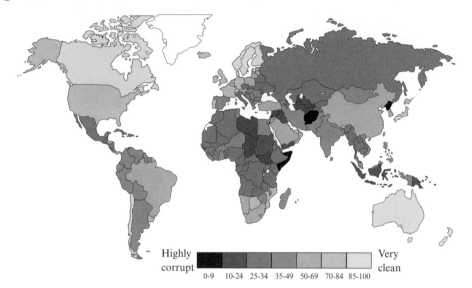

Highly corrupt ▮▮▮▮▮▮▮ Very clean
0-9 10-24 25-34 35-49 50-69 70-84 85-100

Figure 1 Corruption Perceptions Index 2013

Suggest how the information shown in **Figure 1** may determine the level of development of a country.

the more corrupt the goverment the slower developing will happen. this could be due to embezelment of funds.

(2 marks)

> **Guided**

3   Explain **one** indicator that can be used to measure the level of development of a country.

> Remember that **explain** means you need to give a reason.

GDP is a measure of how wealthy a country is: more

wealth means higher rate of development because there are more funds

to improve qaulity of life.

(3 marks)

Development differences

1   Which of the following is the definition of maternal mortality rate?

☒ **A**   The number of women per 1000 who die while pregnant or after childbirth ✓

☐ **B**   The number of children in a country who grow up without a mother

☐ **C**   The number of single mothers in a country

☐ **D**   The average number of children women give birth to in a country   **(1 mark)**

2   Define the term 'death rate'.

the number of deaths per thousand a year ✓   **(1 mark)**

> **Guided**

3   Analyse the data in **Figure 1**.

Country	GDP per capita (US$, 2015)	Fertility rate	School life expectancy
Italy	35 800	1.43	16 years
Brazil	15 800	1.77	14 years
Malawi	1200	5.60	11 years

Figure 1 Data for three countries at different levels of development

(i)   Describe the relationship between the wealth of a country and its fertility rate.

The wealthier a country with a higher GDP per capita, the lower the
fertility rate. Italy has a low fertility rate ✓ and a high GDP along ✓
with a high school life expectancy.   **(2 marks)**

(ii) Explain why fertility rates vary between the countries in **Figure 1**.

Acess to healthcare and medication or contraception, lots of people in malawi
might not have condoms or pills to prevent pregnancy.   **(3 marks)**

4   State **two** differences between the population structures in developed and developing or emerging countries.

> Take care to show two clear differences, rather than just repeating the same point in reverse (for example, X is bigger but Y is smaller is one difference, not two).

Developing countries have larger birth rates ✓ while the life expectancy
is higher in developed countries ✓   **(2 marks)**

Theories of development

1   Study **Figure 1**.

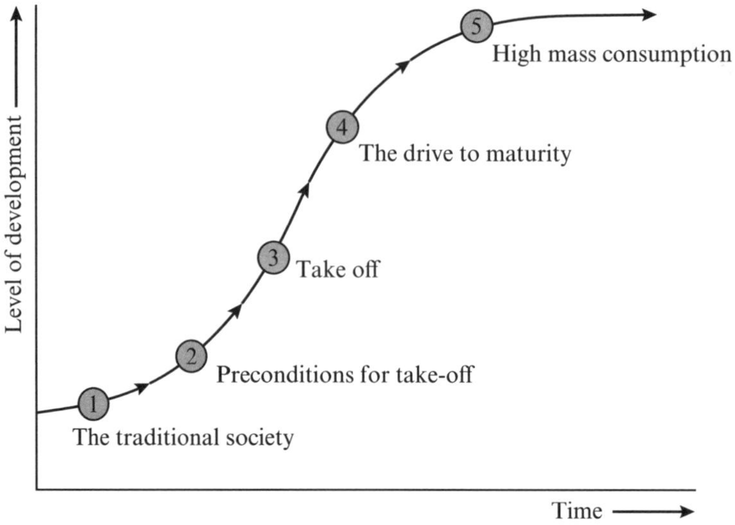

Figure 1 Rostow's modernisation and development theory

Describe how Rostow's theory can be used to help understand economic growth over time.

The further along the model the country is the larger the economic activity.

....

....

(3 marks)

2   Describe how Frank's dependency theory can be used to help explain how countries develop.

....

....

(2 marks)

3   Explain some of the problems with development theories.

....

....

....

....

(3 marks)

Guided   4   Suggest why Frank's dependency theory may not apply to some countries today.

> **Suggest** requires you to give a reasoned explanation and to support your point with an example

Frank's dependency theory was written in the

1930s and does not take into account *globalisation being far*

easier to achieve due to the internet and trade, this means countries

develop a lot quicker.

....

(3 marks)

18

Types of development

1   Which of the following is a characteristic of bottom-up development?

     ☐   **A**   Large-scale government-funded projects

     ☐   **B**   Large-scale projects in rural areas funded by charities

     ☒   **C**   Small-scale projects run by local people

     ☐   **D**   Projects run by TNCs   **(1 mark)**

2   State **two** characteristics of top-down development.

     ....

     ....

     ....

     .... **(2 marks)**

> **Guided**

3   Explain **two** differences between top-down and bottom-up development projects.

     (i)   Top-down schemes are large-scale and aimed at helping large

     numbers of people, or even the whole country, whereas ....

     ....

     ....

     (ii)   Bottom-up schemes are more affordable, using appropriate

     technology that ....

     ....

     .... **(4 marks)**

4   Explain the positive impacts globalisation has had on different groups of people.

> If a question asks about a group of people, be specific: men, women, business people, children, workers, consumers.

     ....

     ....

     ....

     ....

     ....

     ....

     .... **(4 marks)**

Approaches to development

1 Study **Figure 1**.

Figure 1 The Three Gorges Dam, a hydroelectric dam on the Yangtze River in China

Suggest **two** disadvantages of top-down projects such as the one shown in **Figure 1**.

....

....

....

.... **(3 marks)**

2 Define the term 'intermediate technology'.

..... technology that is suitable for the current stage of development ....

.... **(1 mark)**

> **Guided**

3 Explain **one** advantage of intermediate technology.

Intermediate technology is affordable for small rural communities. It

makes use of local ..... skills ....

....

.... **(3 marks)**

4 Many TNCs have located their operations in emerging countries. Suggest how people in emerging countries can benefit from TNCs.

> Be careful to only discuss the benefits, and try to be specific.

....

....

....

....

.... **(3 marks)**

🌐 Case study **Location and context**

1 Describe the location of the emerging country that you studied in your case study.

....

....

....   **(3 marks)**

> Try to include: the continent, neighbouring countries, main physical and human features such as mountains, rivers, coasts and cities, and transport links.

Guided ▷

2 Explain **one** feature of the political context of the emerging country you have studied.

> Consider trade with neighbouring countries and others, ports and natural resources.

India imports oil and other raw materials from China and Saudi Arabia

and exports ....

....

....

....   **(3 marks)**

3 Study **Figure 1**.

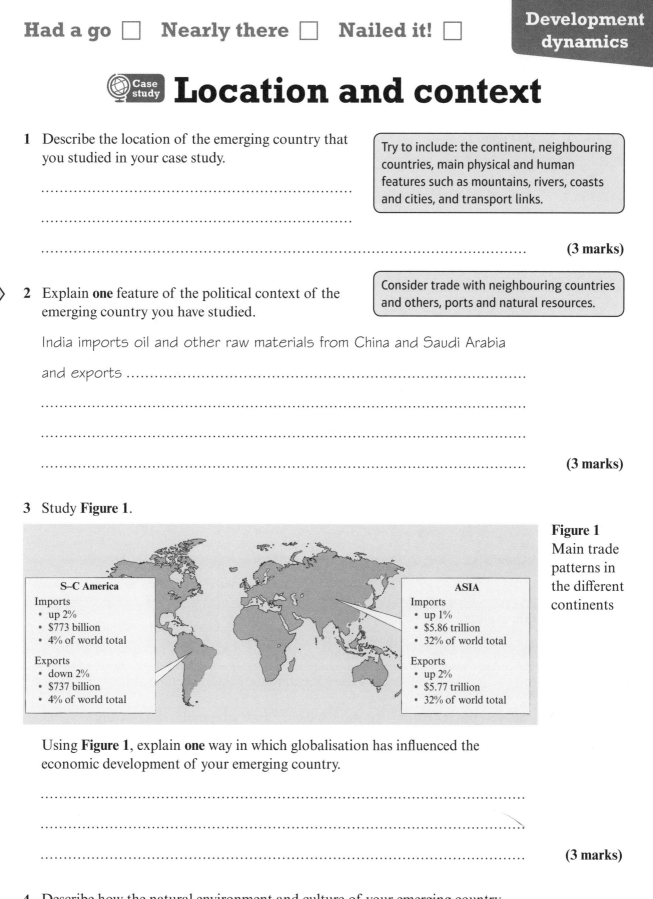

S–C America
Imports
- up 2%
- $773 billion
- 4% of world total

Exports
- down 2%
- $737 billion
- 4% of world total

ASIA
Imports
- up 1%
- $5.86 trillion
- 32% of world total

Exports
- up 2%
- $5.77 trillion
- 32% of world total

Figure 1
Main trade patterns in the different continents

Using **Figure 1**, explain **one** way in which globalisation has influenced the economic development of your emerging country.

....

....

....   **(3 marks)**

4 Describe how the natural environment and culture of your emerging country case study has influenced its economic development.

....

....

....   **(3 marks)**

> Consider the positives and negatives. Remember that culture refers to the way of life in that country. Do the natural environment and culture attract tourists? Is the environment difficult to live in? Does the culture restrict economic development?

Case study Globalisation and change

1 Study **Figure 1**.

You need to have numerical economic data for the emerging country you have studied. You can find data online from reliable sources such as the CIA's World Factbook or the World Bank.

Economic indicator	India	Your case study
GDP per capita (US$)	5800	
GNI per capita (US$)	5630	
Economic employment sectors		
Primary (%)	49	
Secondary (%)	20	
Tertiary (%)	31	
Main imports	Crude oil, precious stones, machinery, chemicals	
Main exports	Petroleum products, vehicles, machinery, iron and steel, pharmaceuticals, cereals	

Figure 1 Economic data for India

(i) Suggest **two** indicators that show India (or your case study) has recently experienced economic development.

....

.... (2 marks)

(ii) Give one piece of evidence which suggests that India is an emerging country.

....

.... (2 marks)

Guided 2 Explain **one** indicator that shows that India is not a developed country.

The high percentage (almost half) of people still employed in ....

....

.... (2 marks)

🌐 Case study **Economic development**

1   Study **Figure 1**.

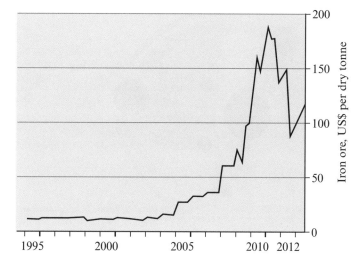

Figure 1 The cost of iron ore, 1995–2012

The peak is when the prices were at their highest. Remember to always give the units.

(i)   In which year did the price of iron ore peak?

.... 2010 .... **(1 mark)**

(ii) Calculate the difference between the peak price of iron ore and its price in 2012.

.... 190 - 90 = 100   ✗   ₹ 10 .... **(1 mark)**

2   Suggest how TNCs may be affected by the changes in iron prices shown in **Figure 1**.

....

.... **(2 marks)**

Guided ⟩ 3   Explain the impact of globalisation on the environment of emerging countries.

TNC factories may cause air and water pollution in emerging countries.

Transporting ....

....

.... **(3 marks)**

4   Referring to a named developing or emerging country, describe how levels of development vary within a country.

....

.... **(2 marks)**

🌐 Case study International relationships

1   Study **Figure 1**.

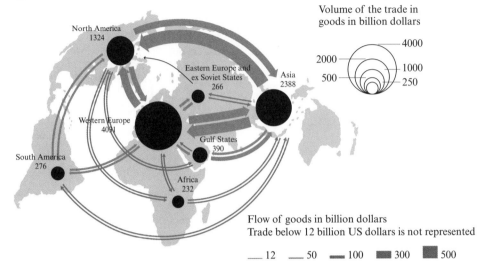

Figure 1 A flow map showing volumes of global trade

What is the largest trading area?

☐ **A**   North America   ☒ **C**   Western Europe

☐ **B**   Asia   ☐ **D**   South America   **(1 mark)**

2   Which of the following is the value of Asia's exports to North America?

☐ **A**   50 billion dollars   ☐ **C**   300 billion dollars

☐ **B**   100 billion dollars   ☒ **D**   500 billion dollars   **(1 mark)**

⟩ **Guided** ⟩   3   For a named emerging country, explain how global trade has contributed to the economic development of the country.

> You need to be specific here and name trading partners and the types of goods traded.

Trade allows emerging countries to import raw materials

and energy resources such as ....

....

....

....   **(3 marks)**

4   Explain **one** way international organisations have affected the development of the economy of an emerging country you have studied.

> Refer to organisations such as the WTO, the World Bank and the IMF.

....

....

....

....   **(3 marks)**

Case study **Costs and benefits**

1   For a named emerging country, assess to what extent globalisation
    has had a positive impact on the quality of life for the people living there.

> - Discuss the main TNCs, trading partners and transport routes of your named country.
> - Think about whether trade and TNCs have had positive impacts on the people, and describe them. Think about the types of jobs and skills created, income levels and so on.
> - Consider whether any groups or areas have not benefited, or have been negatively impacted.
> - Use facts and figures to support your points. For example, give the changing GDP per capita figures and state any regional differences.
> - End with a conclusion, stating the most important factors to support whether you think globalisation has had a positive impact for the people in your named country.
> - Use key geographical terms and paragraphs to structure your answer.
> - Check your work. Remember that four marks will be awarded for spelling, punctuation, grammar and specialist terminology.

Named emerging country: ....

....

....

....

....

....

....

....

....

....

....

....

....

....

....

....

....

(8 marks – plus 4 marks for SPGST)

> In your exam you should write approximately one side of extended
> writing for an 8-mark question. Finish off your answer on your own paper.

Urbanisation trends

1 Define the term 'urbanisation'.

.....The movement of ....people towards an urban area...
.... **(1 mark)**

2 Study **Figure 1**.

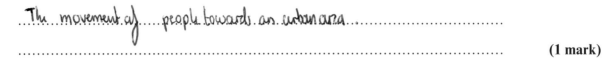

Figure 1 Urbanisation trends in major regions of the world, 1950–2050

In the year 2000, what percentage of the world's population lived in urban areas?

☐ **A** 38 per cent   ☐ **C** 70 per cent

☒ **B** 48 per cent   ☐ **D** 80 per cent   **(1 mark)**

3 What is the likely projected urban population increase for Africa between 1950 and 2050?

☐ **A** 12 per cent   ☐ **C** 53 per cent

☒ **B** 46 per cent   ☐ **D** 88 per cent   **(1 mark)**

⟩ **Guided** ⟩ 4 Using the information in **Figure 1**, compare the projected urbanisation trends of the developed world with the developing world.

> When comparing use linking words such as whereas, on the other hand, in contrast.

In North America and Europe (developed world) the rate

of urbanisation is projected to slow down, whereas ....

....

....

.... **(2 marks)**

Megacities

1   Which statement best describes a megacity?

   ☐ **A**   A city where the population is more than 5 million

   ☐ **B**   A city with a population of 8–10 million

   ☒ **C**   A city with a population of more than 10 million

   ☐ **D**   A city with a population of more than 20 million   **(1 mark)**

> **Guided**

2   Study **Figure 1**.

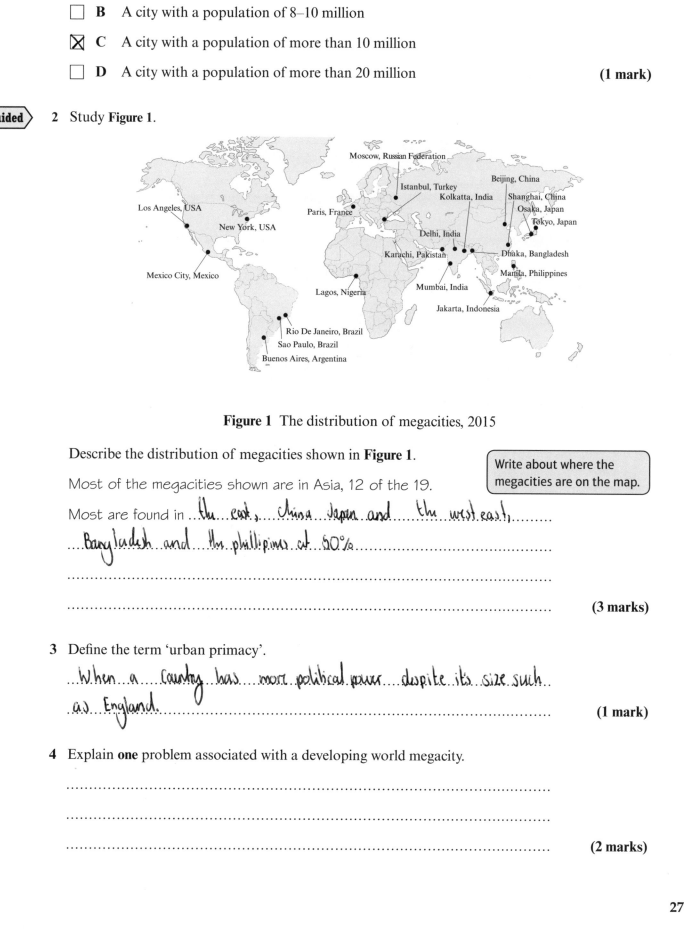

Figure 1 The distribution of megacities, 2015

Describe the distribution of megacities shown in **Figure 1**.

> Write about where the megacities are on the map.

Most of the megacities shown are in Asia, 12 of the 19.

Most are found in ...the east, china Japen and the west east,......

...Bangladesh and the phillipines at 50%.....................................

....

....   **(3 marks)**

3   Define the term 'urban primacy'.

...When a country has more political power despite it's size such...

...as England...   **(1 mark)**

4   Explain **one** problem associated with a developing world megacity.

....

....

....   **(2 marks)**

Urbanisation processes

1   Study **Figure 1**.

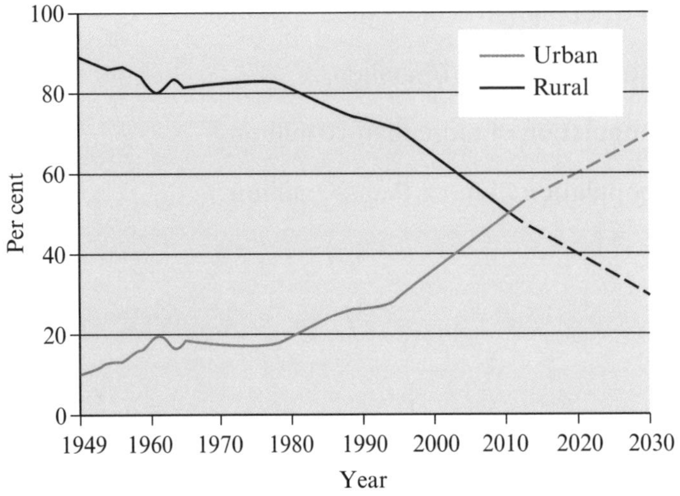

Figure 1 China's population change

(i)   In which year were there equal numbers of people living in rural and urban areas in China?

☐ **A**   1980   ☐ **C**   2000

☐ **B**   1990   ☒ **D**   2010   **(1 mark)**

(ii)   Using **Figure 1**, describe how urbanisation has increased between 1980 and 2015.

....The urbanisation rate of growth has increased from 1980........
to 2015...
.... **(2 marks)**

> **Guided**   2   Explain **two** reasons for the changes shown between 1980 and 2015.

Use data and figures from the information provided. Take care not to simply reverse and repeat the same point.

Lack of well-paid jobs in agriculture in rural areas causes people, mainly

the young, to ....

....

....

....

.... **(4 marks)**

3   Explain **one** reason why some cities in the developed world have experienced a decline in population.

....

....

.... **(2 marks)**

Differing urban economies

1   Define the term 'informal employment'.

...the employment not under const contract. This usually means... ...non-set hours and,...   **(1 mark)**

2   Study **Figure 1**.

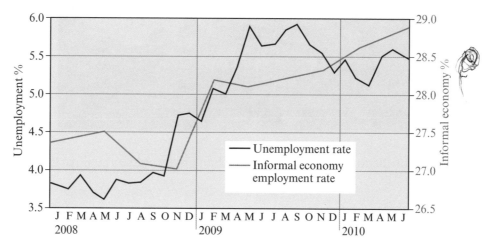

Figure 1 Unemployment and informal economy rates in Mexico

Identify the four-month period in which the informal economy grew fastest.

.....October November December January.....   **(1 mark)**

> **Guided**

3   Explain **one** problem for people working in the informal economy.

People who work in the informal sector do jobs that are not officially

recognised and therefore workers ....

....

....

....   **(3 marks)**

4   Compare the economic activities in developed countries with those in developing or emerging countries.

> You need to refer to the primary, secondary, tertiary and quaternary economic sectors and the levels of skill and pay involved in each.

....

....

....

....

....   **(3 marks)**

Changing cities

1   Study **Figure 1**.

Figure 1 An aerial photograph of Kingswood Retail Park, near Hull.

In which of the following parts of the urban area would a retail park most likely be found?

☐   **A**   CBD

☐   **B**   Suburbs

☐   **C**   Inner city

☐   **D**   Rural–urban fringe   **(1 mark)**

▷ **Guided** ▷   2   Describe **two** different land uses, other than retail, shown in **Figure 1**.

(i)   The large, rectangular buildings indicate that factories could be

     in this area.

(ii) ...There are lots of house, this could suggest residential

...land use.

                                         (2 marks)

3   Explain why businesses and industries choose to locate in areas like that shown in **Figure 1**.

> Use evidence from the image to help you.

                                                       (4 marks)

🌐 Case study **Location and structure**

Guided 1 You have studied a megacity in an emerging or developing country as a case study. Describe your megacity's location.

> Try to include: where in the country the megacity is, how well connected it is to other places, its transport links.

Mumbai is a megacity in the Deccan region of India.

Mumbai lies on the western coast ....

....

.... **(3 marks)**

2 Explain **one** reason why the location of the megacity you studied has been important for its growth.

> Does it have a coastal location that has enabled a port to develop and trade? Does it have certain resources that mean particular industries have grown and attracted workers?

....

....

....

....

.... **(2 marks)**

3 Study **Figure 1**.

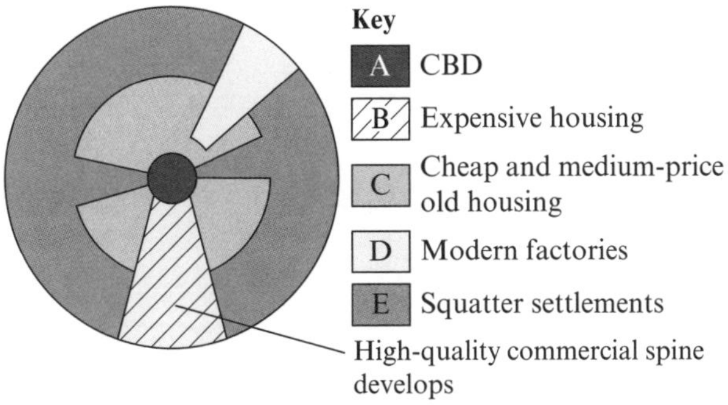

Key
- A CBD
- B Expensive housing
- C Cheap and medium-price old housing
- D Modern factories
- E Squatter settlements
- High-quality commercial spine develops

Figure 1 A land use model

Which of the following best describes the land use in zone A?

☐ **A** Low-quality squatter settlements

☐ **B** Heavy industry

☒ **C** Large shops and commercial offices

☐ **D** High-class residential **(1 mark)**

4 Referring to a megacity you have studied, describe the land use in and around the CBD.

> Be specific here: name businesses and areas.

....

....

.... **(2 marks)**

🌐 Case study Megacity growth

1 Study **Figure 1**.

Mumbai
12.6 million   1992   16.4 million   2001   18.4 million   2011

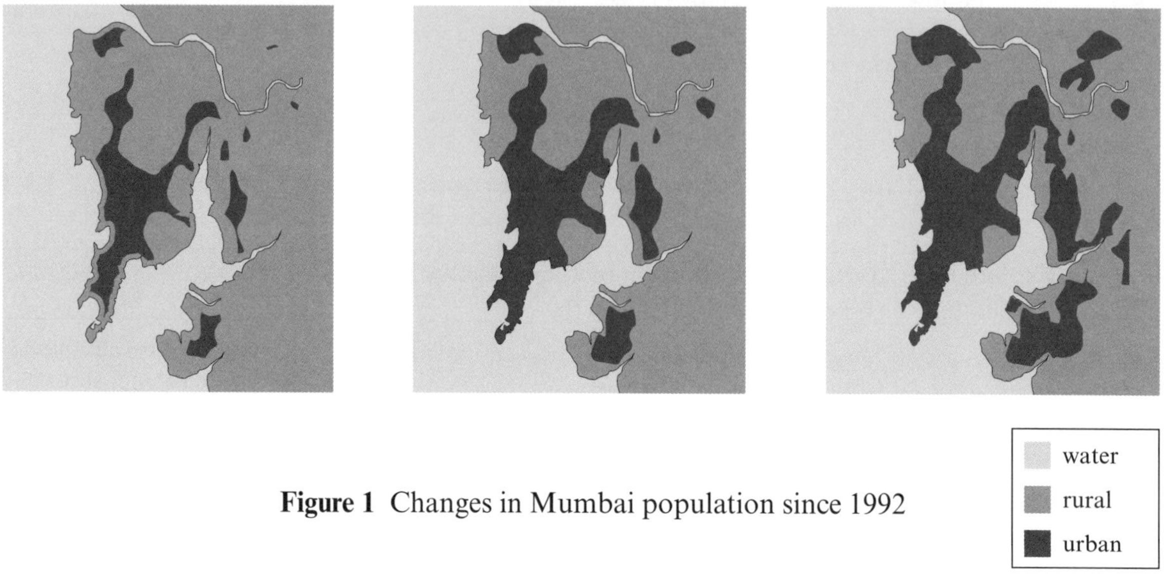

☐	water
▨	rural
■	urban

Figure 1 Changes in Mumbai population since 1992

Describe the changes that have occurred in Mumbai population since 1992.

There has been an increase in urbanisation, meaning the amount of urban areas decreased.

(2 marks)

2 Referring to a megacity you have studied, suggest **two** reasons why megacities grow.

Refer to specific examples and give facts and figures.

....

....

....

....

....

(4 marks)

Guided ▷   3 Referring to a megacity you have studied, explain how it has changed as a result of recent growth.

As the population of a megacity increases, as more and more people

move into the city, more houses and services will be built ....

....

....

(3 marks)

🌐 Case study **Megacity challenges**

1   Referring to the megacity you studied, explain **one** social problem facing people living there.

> Be specific, name places and give facts and figures.

....

....

....   **(3 marks)**

Guided ▷   2   Explain **one** reason why large numbers of people in megacities in developing and emerging countries work in the informal sector.

As more people move into megacities, the demand for jobs is high. Not

all migrants may have skills and therefore ....

....

....

....   **(3 marks)**

3   Study **Figure 1**.

Figure 1 Residential areas in a megacity

(i)   Compare the different living conditions shown in different parts of **Figure 1**.

....

....

....

....   **(3 marks)**

(ii)   Explain **one** reason for the differences in living conditions in megacities.

....

....

....   **(2 marks)**

33

🌐 Case study **Megacity living**

1 For a named megacity in a developing or emerging country, compare the residential areas of different groups of people.

....

....

....

.... **(3 marks)**

2 Explain **one** reason why there are inequalities in how people are housed in megacities.

....

....

....

.... **(3 marks)**

Guided **3** Explain **one** advantage of a top-down strategy used to make a megacity more sustainable.

One advantage of top-down strategies is that they are large-scale, so

....

....

.... **(2 marks)**

4 Explain **one** disadvantage of a bottom-up strategy used to make a megacity more sustainable.

....

....

.... **(2 marks)**

5 Referring to a named example, explain how living conditions in squatter settlements can be improved.

> Improvements can be made on a small scale by self-help schemes.

....

....

....

.... **(3 marks)**

Case study **Megacity management**

1 For a named megacity, evaluate the attempts made to improve the quality of life for people living there.

> To answer this question:
> - break the question down
> - think about what has been done to improve the quality of life for people, then describe specific schemes – aim to describe two
> - address how sustainable these schemes have been – the impact on people, the environment and the economy
> - discuss the advantages and disadvantages of the schemes, how different groups of people have been affected and suggest alternatives
> - make a judgement on whether the attempts have been successful
> - summarise your ideas in a short conclusion.

Named megacity: ....

....

....

....

....

....

....

....

....

....

....

....

....

....

....

....

....

....

....

....   **(8 marks)**

> In your exam you should write approximately one side of extended writing for an 8-mark question. Finish off your answer on your own paper.

Paper 1

1 For an emerging country you have studied, assess the social and economic impact of government policy.

> For this question you need to explain your understanding of the impact of government policy for your emerging country case study.
>
> To address Assessment Objective 2, which is worth 4 marks, you could explain how government policies impact on people and the economy. Consider:
> - Trade policy – with the EU and USA and other parts of the world. Free trade or protectionist?
> - Privatisation policy – have national industries such as oil and gas been allowed to be taken over by private companies?
> - Foreign direct investment – have TNCs been allowed to open and set up operations in the country?
> - Health and education policy – is there investment in research and science?
>
> To address Assessment Objective 3, which is also worth 4 marks, remember to:
> - Weigh up the positive and negative impacts of the policies outlined on the **economy** and **people** of the country you have studied.
> - Come to a judgement about what is the most important point and support this with evidence (with facts and figures).
>
> This question has 4 marks for SPGST, so remember to:
> - spell and punctuate with consistent accuracy
> - use the rules of grammar with effective control
> - use a wide range of specialist geographical terms.

....

....

....

....

....

....

....

....

....

....

....

....

(8 marks + 4 marks for SPGST)

> Continue your answer on your own paper. You should aim to write approximately one side of A4.

Uplands and lowlands

> **Guided**

1 Study **Figure 1**.

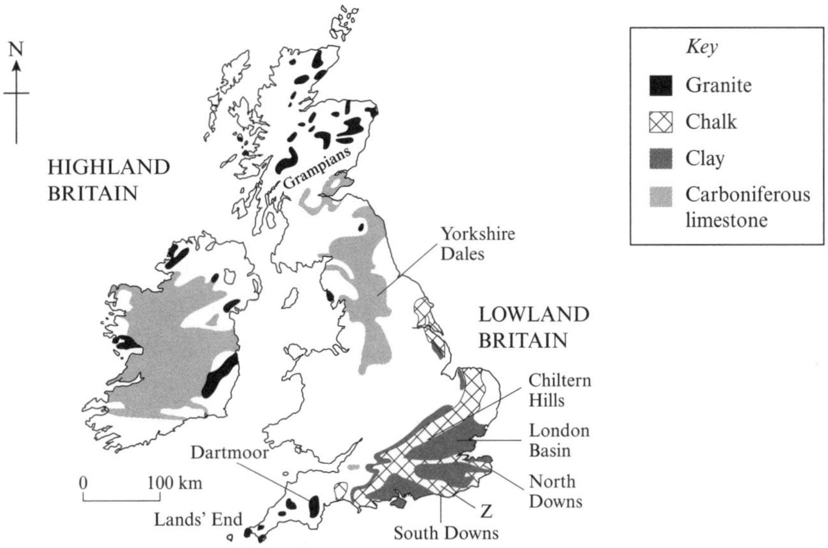

Figure 1 A map showing some of the rock types of the British Isles

(i) Which of the following best describes area Z?

☐ **A** An upland area made up of mainly sedimentary rocks

☐ **B** A lowland area made up of mainly igneous rocks

☒ **C** A lowland area made up of mainly sedimentary rocks

☐ **D** An upland area made up of mainly igneous rocks   **(1 mark)**

(ii) Explain **one** way past tectonic processes have caused the development of the Grampian mountains in Scotland.

Many of the UK's upland mountains are ancient volcanoes, formed from

...Tectonic convergence., the plates smashed into each other smashing the land...

....upwards:..

....   **(2 marks)**

2 Explain **one** way glacial processes have contributed to the formation of the UK's physical landscape.

...Glaciers melted clawing glacial till that scraped through the ground...........

.....leaving behind   U and V-shaped valleys...................................

....   **(2 marks)**

> Glaciers once eroded the landscape and carved distinct features found in parts of the UK today.
> Glaciers also deposited large amounts of material, creating much of the lowlands of the UK.

Main UK rock types

1 Identify which of the following is an igneous rock.

☒ **A** Granite

☐ **B** Chalk

☐ **C** Carboniferous limestone

☐ **D** Clay

(1 mark)

2 State the main characteristics of a sedimentary rock.

> When describing characteristics of a rock, try to refer to what it is made up of, its colour and main features.

..... They are ; soft, pores and layerd ....

....

(2 marks)

Guided

3 Slate is an example of a metamorphic rock. Explain how metamorphic rocks form.

Slate is formed from clay. Layers of clay are compressed under ....

... Heat and pressure ....

....

(2 marks)

4 Study **Figure 1**.

Figure 1 Great Mis Tor, a granite rock formation, Dartmoor

> Use all the information in the caption. It tells you this landform is made of granite.

Explain how this feature was formed.

....

....

....

(2 marks)

Physical processes

1 Study **Figure 1**.

Figure 1 Cwm Cau on the peak of Cadair Idris, Snowdonia National Park, Wales

Which of the following describes the feature shown in **Figure 1**?

☐ **A** A U-shaped valley formed by a glacier

☒ **B** A corrie formed by glacial erosion

☐ **C** A ribbon lake formed from glacial melt water

☐ **D** A delta formed from river deposits   **(1 mark)**

> **Guided**

2 Suggest **one** way this feature has been modified by recent weathering.

Since the glaciers melted, the rocks in this area will have been

weathered by ..... Freze-thaw weathering due to the water having gan

..... in the cracks of the rocks .....

.....

.....   **(2 marks)**

3 Study **Figure 2**.

Figure 2
A V-shaped valley

Suggest **one** post-glacial process that has formed the valley shown in **Figure 2**.

> Use information from the photo. How has the river in the centre of the photo contributed to the formation of the valley?

...... V-shaped valley → glaciation .....

.....

.....   **(2 marks)**

Had a go ☐   Nearly there ☐   Nailed it! ☐

Human activity

1   Study **Figure 1**.

Figure 1 A view of Grasmere in the Lake District

> Woodland would once have covered this landscape. What can you see where there is no woodland?

(i)   State **two** ways humans have modified the landscape shown in **Figure 1**.

....

....

....   **(2 marks)**

(ii) Explain why the landscape in **Figure 1** may not be suitable for arable farming.

This is an upland mountainous area. The slopes are ....

....

....   **(2 marks)**

2   Study **Figure 2**.

Figure 2 Fowey in Cornwall

Explain **one** reason why this landscape was chosen for a settlement.

....There is a large surrounding body of water which allows travel and trade...   **(2 marks)**

40

Geology of coasts

1 Define the term 'concordant coast'.

...... Bands of hard and soft rock run parallel to the coast .... **(1 mark)**

> **Guided** > **2** State **two** differences between **discordant** and **concordant** coasts.

A discordant coast is where alternate bands of hard and soft rock run

at right angles to the coast, whereas ....

.... **(2 marks)**

3 Describe **two** characteristics of soft rock cliffs.

....

.... **(2 marks)**

4 (i) Study **Figure 1**.

Year	2005	2006	2007	2008	2009	2010	2011	2012	2013	2014
North Barmston	0.3	0.0	4.48	0.0	0.0	0.0	0.0	0.0	0.96	0.0
South Barmston	2.29	0.31	3.77	1.03	1.37	2.4	0.49	3.63	0.0	0.0

Figure 1 Erosion (metres lost each year) along parts of the East Yorkshire coast

Calculate the mean rate of erosion for South Barmston to **two** decimal places.

> To calculate the mean you need to **add up** all the values in the set and then **divide** the total by the number of values. In this case there are 10 values.

....

.... **(2 marks)**

(ii) Study **Figure 2**.

Figure 2 Aerial view of Barmston

— caravan park

— rock groyne

Suggest why erosion is greatest to the south of the caravan park in Barmston.

...... Because the direction of waves is from south to North ....

.... **(2 marks)**

Landscapes of erosion

1 Study **Figure 1**.

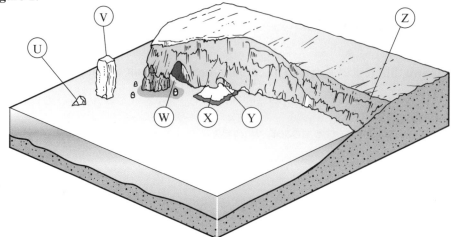

Figure 1 A range of different landforms of coastal erosion

(i) Complete the following table by writing the letters **U, V, W, X, Y** or **Z** into the boxes to match each landform with the correct label on **Figure 1**. The first one has been done for you.

Landform	Stump	Wave-cut platform	Cliff	Stack	Arch	Wave-cut notch
Letter	U	X	Z	V	W	Y

(4 marks)

(ii) Suggest how **one** landform in **Figure 1** is formed.

..~~An arch is~~.. A stack is formed due to the
lack of structural support in an arch. The
arched rock collapses leaving behind a stack. ....

....

> You will need to refer to coastal erosion processes and sub-aerial weathering, and the sequence of events that have formed this landform.

(3 marks)

> **Guided** 2 Add two further labelled diagrams to show how a wave-cut platform is formed.

New rock face exposed
Area attacked by waves between high and low water marks
HW
LW
Sea

(4 marks)

Waves and climate

1 Which of the following is an example of how waves erode the coast?

☒ **A** Longshore drift

☐ **B** Abrasion

☐ **C** Slumping

☐ **D** Deposition **(1 mark)**

2 Explain **two** differences between destructive and constructive waves.

> Remember to make use of comparative connectives such as 'whereas' and 'however'.

...Constructive waves deposit sediment onto the beach and contain swash...
...whereas destructive waves do pull out sediment with backwash,...
...more powerful retraction..

(2 marks)

Guided

3 Describe **one** way in which wave action can erode coastal cliffs.

> Think about abrasion and attrition.

Waves strike the base of the cliffs. The hydraulic action of

the water wears the rock in the cliff face away bit by bit through the

process of erosion. This repeated action of waves striking the base of

the cliff creates a wave-cut notch at the point where ....

....

....

.... **(3 marks)**

4 State **two** factors that influence the size and type of wave.

....

....

....

.... **(3 marks)**

5 State the type of waves that would be associated with a high pressure weather system (clear skies and low winds).

....

.... **(1 mark)**

Sub-aerial processes

> **Guided**

1 Study **Figure 1**.

Figure 1 Mass movement on the western end of Higher Dunscombe Cliff, Dorset

(i) Which of the following processes best explains what has happened in **Figure 1**?

> Don't rush the multiple-choice questions. Consider all the options carefully.

☐ **A**   Quarrying

☐ **B**   A landslide

☐ **C**   Abrasion

☒ **D**   Soil creep   **(1 mark)**

(ii) Suggest **one** factor that could explain the mass movement in **Figure 1**.

Heavy rainfall will have caused the cliffs to become saturated .meaning....

...the ground is much softer and rocks can move through them more easily. This........

....could also lead to freeze-thaw weathering...................................... **(2 marks)**

2 Define the term 'sub-aerial processes'.

....

....

.... **(2 marks)**

3 Explain **one** reason why coasts retreat at different rates.

....

....

....

.... **(3 marks)**

Transportation and deposition

 Guided 1 Draw a labelled diagram to explain the process of longshore drift.

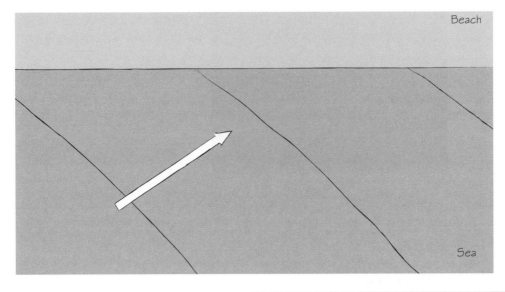

Complete the student's diagram to show how the swash and backwash move material along the beach.

(4 marks)

Guided 2 Study **Figure 1**.

Figure 1
An example of a coastal landform feature

(i) Name the type of landform shown in **Figure 1**.

...... Bay .... **(1 mark)**

(ii) Suggest **one** way this landform is formed.

The landform is made up of eroded material (sand, shingle and stones)

that has been transported from elsewhere and deposited by the sea.

....

....

.... **(3 marks)**

Landscapes of deposition

1 Study **Figure 1**.

Figure 1 A depositional landform off the coast of Helston, Cornwall

| spit bar stack arch |

What is landform A? Choose from the words in the box. A = …Bar……… **(1 mark)**

2 Study **Figure 2**.

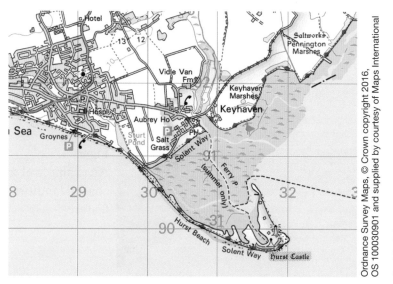

Ordnance Survey Maps, © Crown copyright 2016, OS 100030901 and supplied by courtesy of Maps International

Figure 2 Ordnance Survey map extract showing a coastal landform

(i) Name the type of landform shown in the map.

…………Spit………………………………………………………………… **(1 mark)**

(ii) Draw a labelled diagram to show how this landform is formed.

> Make sure your drawings are clear – it is much more important that the details are clear and accurately labelled than that the drawing is shaded or coloured in. Your labels need to be clear enough to explain the sequence of formation.

(3 marks)

Human impact on coasts

1 Study **Figure 1**.

Figure 1 Easington Gas Terminal on the East Yorkshire coast

(i) Which of the following human activities is **not** shown in **Figure 1**?

☒ **A** Retailing

☐ **B** Heavy industry

☐ **C** Agriculture

☐ **D** Transport

(1 mark)

(ii) Suggest **one** positive impact human activity could have on this area.

> Keep to the focus of the question, which is **one** positive impact.

...Industry creates jobs..

....

(1 mark)

Guided 2 Explain how human activities can affect the coastal landscape.

Settlements have been built close to the coast. The weight of the

buildings has ....

....

....

Humans have constructed sea defences, such as groynes, along some

coasts. This has helped to ....

....

.... (4 marks)

🌐 Located example **Holderness coast**

1   Study **Figure 1**.

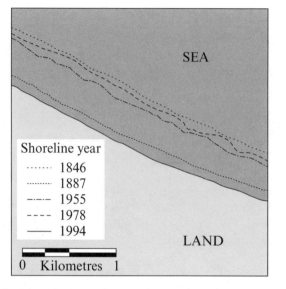

Using the information on the map, measure how much coast has been eroded and calculate over how many years that has happened. To get the mean, divide the total amount of erosion (in metres) by the number of years.

SEA

Shoreline year
...... 1846
········· 1887
–·–· 1955
– – – 1978
——— 1994

LAND

0   Kilometres   1

Figure 1 A diagram showing the rate of coastal recession along a section of the Holderness coast, Yorkshire

Calculate the mean rate of erosion per year on the part of the Holderness coast shown in **Figure 1.** Show your working.

.... **(2 marks)**

2   Suggest possible reasons why the cliff has receded.

....

....

....

.... **(3 marks)**

3   For a named example, explain **one** process that has changed the coastal landscape.

....

....

....

.... **(3 marks)**

You need to refer to any **one** of these: erosion, mass movement, longshore drift or deposition.

⟩ **Guided** ⟩   4   For a named example, explain how coastal management processes have changed the coastal landscape.

Hard engineering has slowed down cliff retreat, and has caused ....

....

....

.... **(3 marks)**

Coastal flooding

1   Study **Figure 1**.

Figure 1 Aerial view of a clifftop caravan site in Norfolk.

Name a land use threatened by coastal flooding.

...... *agricultural* ....   **(1 mark)**

> **Guided**

2   Explain why sea-level rise is threatening coastal areas across the UK.

Sea levels are rising because of climate change. Temperatures have risen,

causing ....

....

....

....   **(4 marks)**

3   Coastal areas are increasingly threatened by rising sea levels. Assess the possible consequences of sea-level rise for people living near the coast.

....

....

....

....

....

....   **(8 marks)**

> This is an extended answer. You need to discuss all the possible consequences of rising sea levels and state which you think is the most likely and most serious. Continue your answer on your own paper. You should aim to write approximately one side of A4.

Coastal management

Guided

1 Explain how groynes can help reduce rates of coastal erosion.

Groynes can help reduce rates of coastal erosion by trapping sediment

that is being moved along the coast by longshore drift. The trapped

sediment ....

....

.... **(2 marks)**

2 Study **Figure 1**.

Figure 1 An example of coastal defences on the south coast of England

Briefly describe how the type of coastal defence shown in **Figure 1** works.

....

....

.... **(2 marks)**

3 Describe **one** disadvantage of beach nourishment.

....

....

....

It is a good idea to revise the advantages **and** disadvantages of a range of coastal management strategies. Cost is often a major concern, but so is how often a process has to be repeated to stay effective.

.... **(2 marks)**

4 Explain **one** advantage of using hard defences at the coast.

....

....

.... **(2 marks)**

5 Explain **one** way Integrated Coastal Zone Management (ICZM) techniques can be used to protect the coastline.

....

.... **(2 marks)**

Investigating coasts: developing enquiry questions

TASK: Investigating the impact of coastal management on coastal processes and communities.

Only work through this page and the two that follow if you did coastal fieldwork. **If you did rivers fieldwork, turn to page 64.**

You will be asked questions on your own fieldwork experiences, but you will also be required to apply the skills you have gained from your fieldwork to some unfamiliar situations.

Study **Figure 1**.

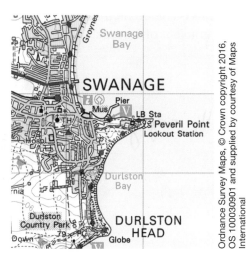

Figure 1 Part of Swanage in Dorset

Ordnance Survey Maps, © Crown copyright 2016, OS 100030901 and supplied by courtesy of Maps International

1   Explain why the area shown in **Figure 1** might be a good place to investigate the impact of coastal management on coastal processes.

   ....

   ....

   ....

   ....   **(3 marks)**

2   State **one** enquiry question relating to the investigation task that could be looked at in this area.

   | Think about and refer to the enquiry questions you devised for your coastal investigation. |

   ....

   ....   **(1 mark)**

⟩ Guided ⟩   3   Describe **one** piece of primary data that could be collected in this area that could help investigate the enquiry question you have stated above.

   The profile of the beach gradient could be measured from ....

   ....

   ....

   ....   **(2 marks)**

Investigating coasts: techniques and methods

1 Describe **one** qualitative fieldwork method you used to investigate the success of a coastal management measure.

> **Qualitative** means describing rather than measuring.

...... Peoples written opinion on how it looks compared to before ......

....

(2 marks)

> **Guided**

2 Study **Figure 1**.

(i) Describe why this is a good method of measuring the gradient of a beach.

It is appropriate to use a clinometer

because the beach has uneven variations

in gradient ...So the ranging pole shows..

..the variation ....

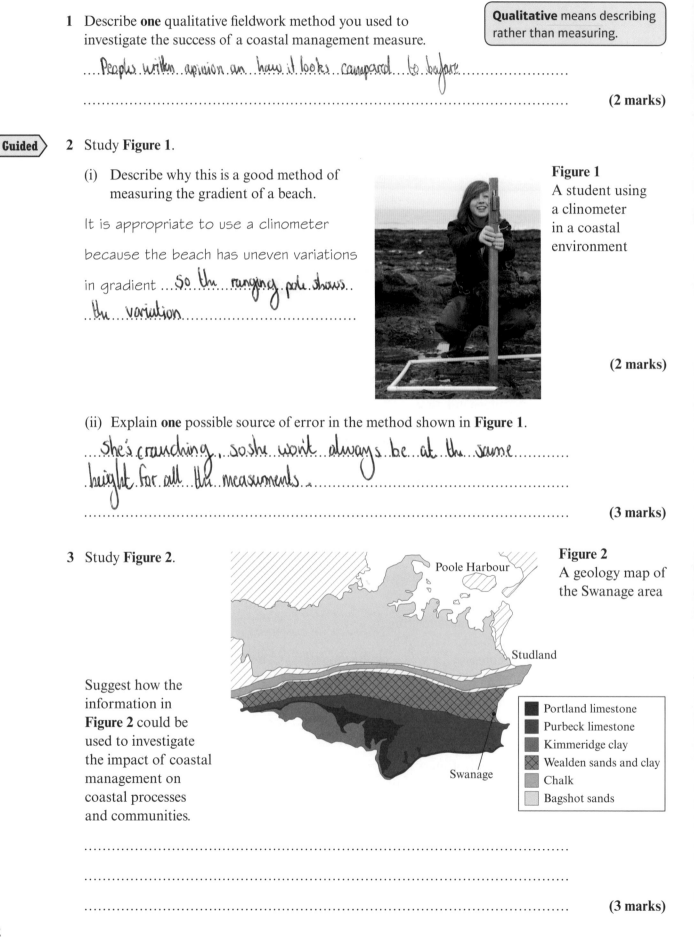

Figure 1
A student using a clinometer in a coastal environment

(2 marks)

(ii) Explain **one** possible source of error in the method shown in **Figure 1**.

...She's crouching, so she won't always be at the same..........

..height for all the measurements. ....

(3 marks)

3 Study **Figure 2**.

Suggest how the information in **Figure 2** could be used to investigate the impact of coastal management on coastal processes and communities.

Poole Harbour

Studland

Swanage

Figure 2
A geology map of the Swanage area

	Portland limestone
	Purbeck limestone
	Kimmeridge clay
	Wealden sands and clay
	Chalk
	Bagshot sands

....

....

....

(3 marks)

Investigating coasts: working with data

1 Study **Figure 1**.

Top of beach	1	2	3	4	5	Sea
Distance between ranging poles	3.2 m	1.5 m	2.2 m	5.1 m	1.6 m	1.2 m
Gradient	−4.5	−7	−5	−2	−9	−4

Figure 1 Data collected by students conducting a coastal investigation in Swanage, Dorset

Suggest **one** presentation technique a student could use to present the data in **Figure 1**.

.... A line graph ....

Think about how you could accurately show the profile of the beach.

.... **(2 marks)**

2 Label **Figure 2** to describe the impact of sea defences in this area.

Figure 2 Sea defences on Swanage beach

(3 marks)

3 Explain **one** way you used GIS in your coastal change and conflict investigation.

....

.... **(2 marks)**

4 Evaluate the suitability of your chosen methods of data presentation in helping you make conclusions about the relationship between coastal management and coastal processes.

Continue your answer on your own paper. You should aim to write approximately one side of A4.

....

....

....

....

....

.... **(8 marks)**

53

River systems

1 Define the term 'long profile'.

....

.... **(2 marks)**

2 Which of the following statements best describes how a river channel changes along its long profile?

☐ **A** The width and depth decrease downstream

☐ **B** The width and depth increase downstream

☐ **C** The gradient increases downstream

☐ **D** The velocity decreases downstream **(1 mark)**

>**Guided**> 3 (i) Complete the diagram by drawing the shape of the channel at points A and C.

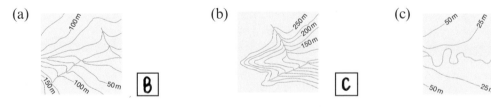

(ii) The contour patterns in (a), (b) and (c) below show what the valley looks like at three different points in the diagram above. Place the letters A, B and C in the box next to each contour pattern to indicate where it would occur on the long profile.

> Contours that are close together indicate a steep slope.

(2 marks)

(a)   ☐ **B**   (b)   ☐ **C**   (c) **A**

>**Guided**> 4 Explain why channel characteristics (discharge, sediment size and shape) change along a river's course at points A and C.

At point A the river is in its upper course. Here the river channel will be

shallow and narrow, the sediment size will be large and angular and

discharge relatively small, whereas at point C ....

....

....

.... **(3 marks)**

Erosion, transportation and deposition

1 Which of the following is the correct definition of hydraulic action?

☐ **A** Rocks hit against each other within the river water and break up

☐ **B** The force of the water on the bed and banks of the river removes material

☐ **C** Scraping away of banks and bed by material in the water

☐ **D** Chemicals dissolve the minerals in the rock   **(1 mark)**

2 Explain the process of abrasion.

> Abrasion involves the load the river is carrying.

....

....

....   **(2 marks)**

> Guided

3 Study **Figure 1**.

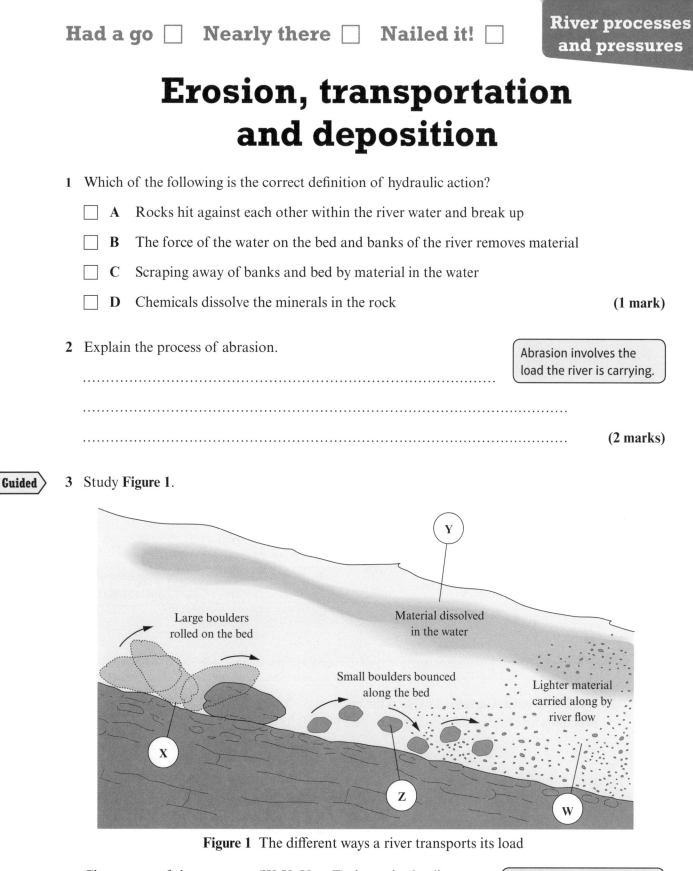

Figure 1 The different ways a river transports its load

Choose one of the processes (W, X, Y or Z) shown in the diagram. Name the process and explain how this process transports material in rivers.

> The large boulders would be transported by traction.

One of the most important processes is traction (X), where ....

....

....

....   **(2 marks)**

Upper course features

1 Study **Figure 1**.

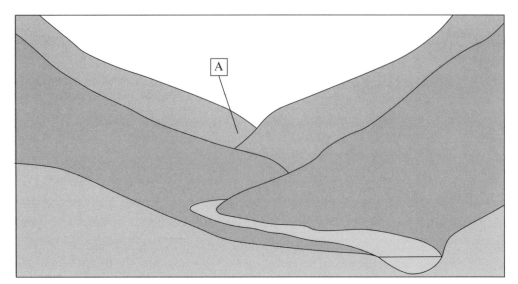

Figure 1 A diagram showing the upper course of a river

(i) Which of the following best describes the feature labelled A?

☐ **A** Meander   ☒ **C** Interlocking spur

☐ **B** Slip-off slope   ☐ **D** Waterfall   **(1 mark)**

(ii) Explain how landform A is formed.

> Your answer needs to include vertical erosion.

....

....

.... **(3 marks)**

>Guided> 2 Draw a labelled diagram to explain the erosion processes involved in the development of a waterfall.

> This diagram has been done for you, but in the exam you would draw it from scratch. You need to add in the labels at **A**, **B**, **C** and **D**.

A

B

D

C

(3 marks)

Lower course features 1

1 Study **Figure 1**.

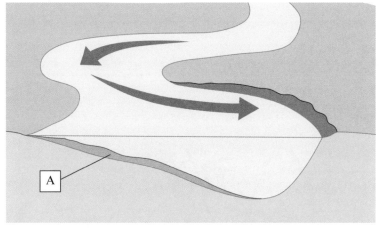

Figure 1 A cross-section through a meander

State the main process that would occur at A on the cross-section diagram.

.... **(1 mark)**

2 Study **Figure 2**.

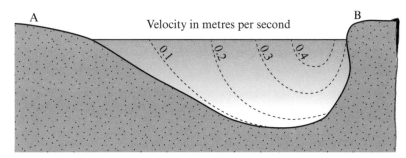

Figure 2 A river meander

Describe the relationship between river depth and velocity on a meander.

> Try to quote figures from the diagram to support your answer.

....The deeper the river the slower the velocity in m/s.....................

....

.... **(2 marks)**

> **Guided**

3 Explain how a meander is formed.

A meander is a bend in a river. Meanders form where the gradient of the

river is less steep. This means that, as a result of a swinging current,

erosion is greatest on the ....right side of the swinging current............

....closes to the insides causing more heavy erosion of the bank inside....

....

.... **(4 marks)**

Lower course features 2

1 Describe the process that forms levées.

....

....

....   **(2 marks)**

2 Study **Figure 1**.

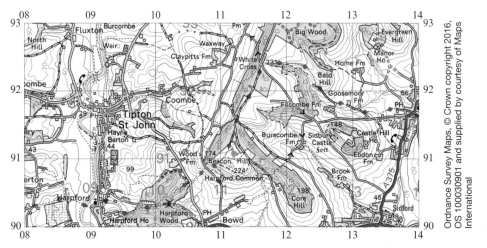

Ordnance Survey Maps, © Crown copyright 2016, OS 100030901 and supplied by courtesy of Maps International

Figure 1
A 1:50 000 scale Ordnance Survey extract showing the River Otter in Devon

Which of the following is the best description of the River Otter to the west of Tipton St John?

> Look at the contours on the map to help you make your choice.

☐ **A** A fast-flowing upland stream

☐ **B** A lowland river flowing through a wide floodplain

☐ **C** A delta close to its mouth

☐ **D** A river flowing over a waterfall in its middle course   **(1 mark)**

3 In **Figure 1**, identify the landform in grid square 08 91.

☐ **A** Interlocking spur   ☐ **C** Meander

☐ **B** Waterfall   ☐ **D** Levée   **(1 mark)**

Guided 4 Explain how deltas form.

Deltas form close to the mouth of a river as it approaches the sea or

lake. The river loses energy and ....

....

....

....   **(3 marks)**

Processes shaping rivers

1 Which of the following is a type of mass movement?

☐ **A** Abrasion   ☒ **C** Slumping

☐ **B** Saltation   ☐ **D** Traction   **(1 mark)**

⟩ **Guided** ⟩ 2 Explain **one** way climate affects river landscapes and sediment load.

The wetter the climate, the higher the river discharge, which will result in

....

....

.... **(3 marks)**

3 Study **Figure 1**.

Figure 1 A river flowing over hard rock

Suggest how the geology in **Figure 1** is influencing the river landforms.

> You need to identify the main river landforms in the picture.

...The..softer..rocks..have..been..eroded..away..leaving..behind..the..hard..rock......

..that..splits..the..river..as..it..flows...

....

.... **(2 marks)**

Storm hydrographs

1   Study **Figure 1**.

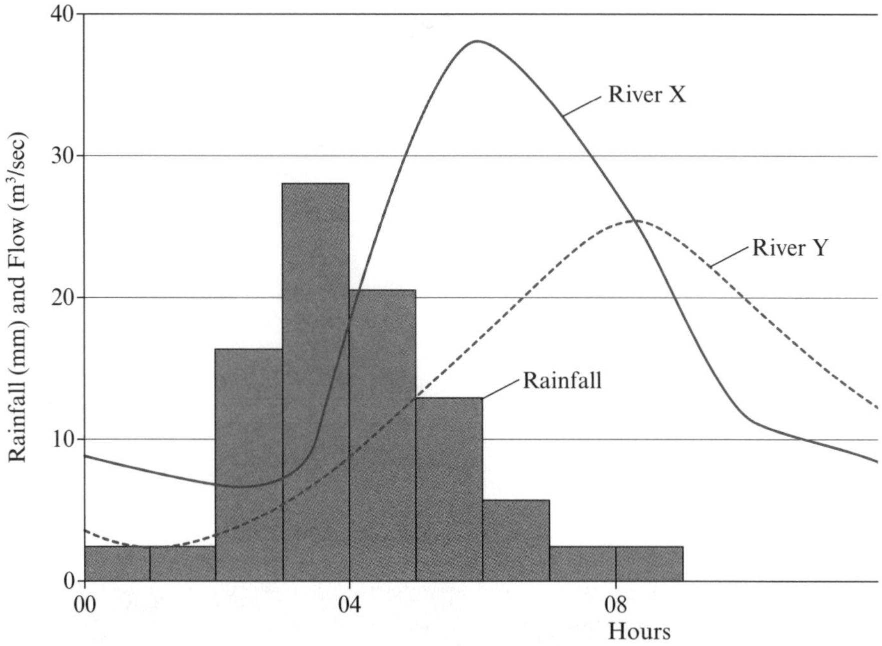

Figure 1 Diagram showing discharge after the same rainstorm in two drainage basins

(i)   Define the term 'rising limb'.

...the increase in discharge from the base flow level until it reaches peak discharge... ✓   **(1 mark)**

(ii) Calculate the lag time for River X.

....2. hours.................................... ✓   ....

| Lag time is the time from the peak rainfall to peak discharge. |

....   **(1 mark)**

(iii) State the peak discharge for River Y.

......25 m³/sec........................ ✓   **(1 mark)**

(iv) Explain **one** reason River X is more likely to flood than River Y.

......Because there is a ....higher. peak discharge, meaning the sea level /........
......ground water level is ....higher... by... 10 cm³/sec..................

....
....   **(3 marks)**

> **Guided**

2   Explain how geology can affect the peak discharge of a river's hydrograph.

Impermeable rocks will not allow water to percolate into the rocks

below the ground. The water then ....

....

....

....

....   **(4 marks)**

🌐 Located example **River flooding**

1   Study **Figure 1**.

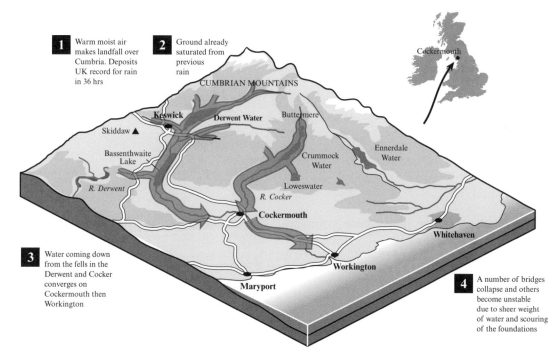

1 Warm moist air makes landfall over Cumbria. Deposits UK record for rain in 36 hrs

2 Ground already saturated from previous rain

CUMBRIAN MOUNTAINS

Keswick   Derwent Water   Buttermere

Skiddaw ▲

Bassenthwaite Lake

R. Derwent

Crummock Water

Loweswater
R. Cocker

Ennerdale Water

Cockermouth

Whitehaven

3 Water coming down from the fells in the Derwent and Cocker converges on Cockermouth then Workington

Workington

Maryport

4 A number of bridges collapse and others become unstable due to sheer weight of water and scouring of the foundations

Cockermouth

Figure 1 Cockermouth in Cumbria, which experienced severe flooding in November 2009

Assess the physical and human factors that contributed to the floods in Cockermouth in 2009.

....

....

....

....

....

....

....

....

....   **(8 marks)**

Assess requires you to discuss all the factors and then decide which is the most important factor that caused the flooding. You need to consider and discuss the contribution of:
- physical factors – the relief of the land, the geology of the underlying rocks, the antecedent weather, the location of the confluence, climate
- human factors – urbanisation, bridges, deforestation.

Continue your answer on your own paper. You should aim to write approximately one side of A4.

Increasing flood risk

1   Which of the following is a threat for the physical landscape if river flooding risk increases?

☐ **A**   More damage to homes

☐ **B**   Loss of food supply as farm land is damaged

☐ **C**   Disruption to transport

☒ **D**   River bank erosion

(1 mark)

> Guided

2   Explain **one** reason river flooding is likely to become more frequent in some parts of the UK.

Climate change may cause temperatures to increase, leading to

evaporation and rainfall; this will cause ....

....

....

....   **(2 marks)**

3   Explain the threats to the environment from more frequent river flooding.

> The focus of the question here is the environmental threats.

....

....

....

....

....

....   **(4 marks)**

4   Suggest **two** economic impacts of more frequent flooding.

....

....

....

....

....

....   **(4 marks)**

Managing flood risk

1 Study **Figure 1**.

Figure 1
A flood defence

(i) Identify the type of hard engineering flood defence shown on **Figure 1**.

....................Dam.. **(1 mark)**

(ii) Explain how this flood defence works.

.........At an area of higher relief the dam holds more water while on.....
.....the other side the relief is lower so the dam keeps more less.......
........water on that side... **(2 marks)**

2 State **one** disadvantage of using soft engineering techniques, such as dredging.

> Soft engineering involves working with natural materials, such as planting trees in the uplands.

....

.... **(1 mark)**

Guided 3 Explain why planting trees in a river catchment helps reduce flooding.

Trees slow down the rate at which water arrives in a river by ....

....

....

....

....

.... **(3 marks)**

Investigating rivers: developing enquiry questions

TASK: Investigating how and why drainage basin and channel characteristics influence flood risk for people and property along a river in the UK.

Only work through this page and the two that follow if you did river fieldwork. **If you did coastal fieldwork, turn to page 51.**

You will be asked questions on your own fieldwork experiences, but you will also be required to apply the skills you have gained from your fieldwork to some unfamiliar situations.

 1 Study **Figure 1**.

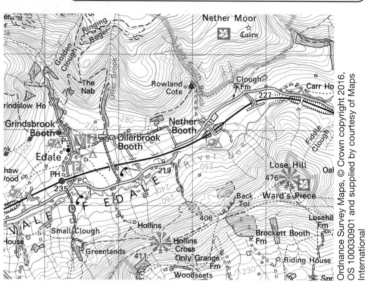

Figure 1 Part of the Vale of Edale in Derbyshire

Ordnance Survey Maps, © Crown copyright 2016, OS 100030901 and supplied by courtesy of Maps International

Explain why the area shown in **Figure 1** might be a good place to investigate how and why drainage basin and channel characteristics influence flood risk for people and property along a river in the UK.

The OS map shows the accessible River Noe and its tributaries in its

upper course, where the river should be shallow enough to take

measurements. ....

....

....

.... **(3 marks)**

2 State **one** enquiry question that you devised to investigate the task.

....

.... **(1 mark)**

3 State **one** piece of primary data that could be collected in this area that could help investigate the enquiry question you have stated above.

....

....

.... **(2 marks)**

Had a go ☐   Nearly there ☐   Nailed it! ☐

Investigating rivers: techniques and methods

1   Which of the following techniques would measure the velocity of a river channel most accurately?

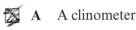 **A**   A clinometer   ☒ **C**   A flow meter

☐ **B**   Time – a float over a certain distance   ☐ **D**   Callipers   **(1 mark)**

> **Guided**

2   Study **Figure 1**.

Figure 1
Collecting data on river channel characteristics

(i)   Suggest **one** reason **Figure 1** shows a good method of measuring river depth.

The tape measure would allow the depth of the river channel to be

....

....   **(2 marks)**

(ii)   Explain **one** possible source of error in the method shown in **Figure 1**.

> Think about the equipment used and how many measurements are taken.

...the clinometer is at an angle...

....

....   **(3 marks)**

3   Study **Figure 2**.

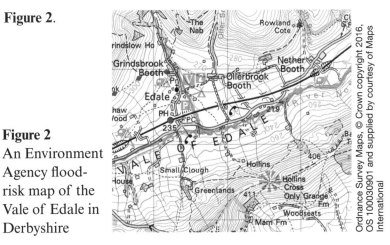

Figure 2
An Environment Agency flood-risk map of the Vale of Edale in Derbyshire

Ordnance Survey Maps, © Crown copyright 2016, OS 100030901 and supplied by courtesy of Maps International

Describe how the information in **Figure 2** could be used to investigate the flooding pressures in this area.

....

....

....   **(3 marks)**

Investigating rivers: working with data

1 Study **Figure 1**.

Distance from water's edge	Site 1 width = 127 cm			Site 2 width = 89 cm		
	Depth (cm)	Pebble length (cm)	Pebble roundness	Depth (cm)	Pebble length (cm)	Pebble roundness
0	0	0	0	0	0	0
25 cm	8	15	angular	23	22	angular
50 cm	19	27	very angular	87	27	angular
75 cm	45	45	sub-angular	73	65	sub-angular
100 cm	38	12	angular			
125 cm	11	8	angular			

Figure 1 Data collected by students conducting a river study in the Vale of Edale area

Suggest **one** data presentation technique a student could use to present the channel width and depth measurements.

> How could you show the profile of the river channel?

....

.... **(2 marks)**

> **Guided**

2 (i) Describe how the cross-sectional area of a river channel is calculated.

To calculate the cross-sectional area, you need to multiply the width by

.... **(2 marks)**

(ii) Calculate the cross-sectional area for site 1 in **Figure 1**.

.... **(1 mark)**

3 Explain **one** way you used GIS in your own river processes and pressure investigation.

....

.... **(2 marks)**

> Continue your answer on your own paper. You should aim to write approximately one side of A4.

4 Assess what techniques you might use to present your data and to help you make conclusions about river drainage characteristics.

....

....

....

....

.... **(8 marks)**

Urban and rural UK

1 Suggest **two** reasons why younger people move to urban areas.

Many young people move out of rural areas to urban areas as there are

more job opportunities, ...

.... **(2 marks)**

2 Study **Figure 1**.

Figure 1 Many rural Post Offices face closure

Which of the following **best** explains why many shops and services in remote rural settlements have closed down?

☐ **A** People prefer to use the services in large urban settlements that are close

☐ **B** The permanent population is declining so fewer people live in these areas to support the services

☐ **C** Tourists bring in their own supplies so do not need to use services when they visit rural settlements

☐ **D** The services in rural areas are generally of a poor quality so people choose not to use them   **(1 mark)**

3 Explain **one** way government policy can reduce the regional differences within the UK.

> Try to explain how Enterprise Zones can help poorer regions.

....

....

.... **(2 marks)**

The UK and migration

1 Study **Figure 1**.

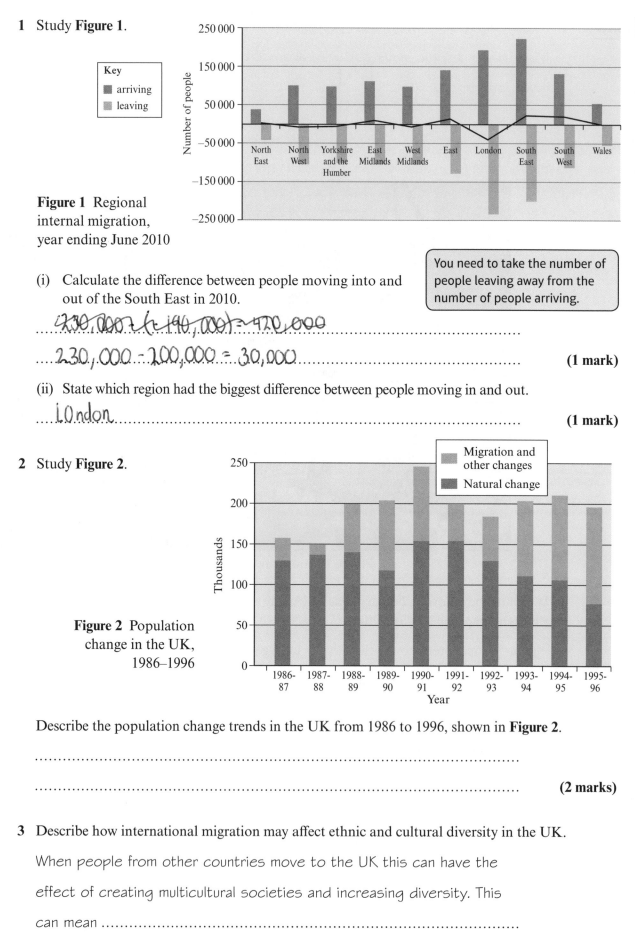

Figure 1 Regional internal migration, year ending June 2010

(i) Calculate the difference between people moving into and out of the South East in 2010.

> You need to take the number of people leaving away from the number of people arriving.

230,000 (-190,000) = 420,000

230,000 - 200,000 = 30,000   **(1 mark)**

(ii) State which region had the biggest difference between people moving in and out.

London   **(1 mark)**

2 Study **Figure 2**.

Figure 2 Population change in the UK, 1986–1996

Key: Migration and other changes / Natural change

Describe the population change trends in the UK from 1986 to 1996, shown in **Figure 2**.

....

.... **(2 marks)**

⟩**Guided**⟩ 3 Describe how international migration may affect ethnic and cultural diversity in the UK.

When people from other countries move to the UK this can have the

effect of creating multicultural societies and increasing diversity. This

can mean ....

.... **(2 marks)**

Economic changes

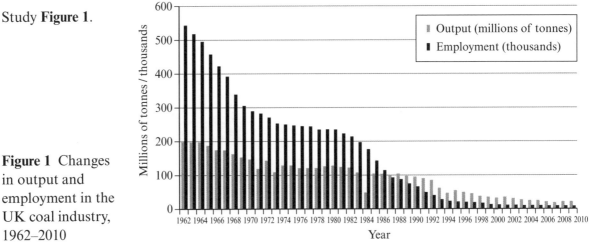

> **Guided**

1 Study **Figure 1**.

Figure 1 Changes in output and employment in the UK coal industry, 1962–2010

Legend:
- Output (millions of tonnes)
- Employment (thousands)

y-axis: Millions of tonnes / thousands (0–600)
x-axis: Year (1962–2010)

Suggest reasons for the decline in the number of coal miners in the UK.

Globalisation has been an important driver to the changes in much

employment in this primary extraction industry, especially the availability

of cheaper sources overseas, e.g. Eastern Europe where labour is

relatively cheap. ....

....

....   **(3 marks)**

2 Study **Figure 2**.

	1973	1979	1981	1990	2010
Agriculture and fishing	1.9	1.6	1.6	1.4	1.5
Manufacturing	34.7	31.3	28.4	20.5	8.2
Distribution, catering and hotels	17.4	18.4	19.1	21.4	21.3
Banking and finance	6.4	7.2	7.9	15.2	20.3

Figure 2 Changes in employment structure of the UK (%), selected types, 1973–2010

(i) Between which years did **secondary industry** see the biggest decline?

> Secondary industry involves making goods in factories.

☐ **A** 1973–1979   ☐ **C** 1981–1990

☐ **B** 1979–1981   ☒ **D** 1990–2010   **(1 mark)**

(ii) Using **Figure 2**, describe the main changes that occurred to the UK's employment structure from 1973 to 2010.

.....Employment change was highest from 1973 to 1981, in 1990-2010 there

.....was the greatest change in employment for Banking and finance

.....while the largest change for the agriculture and manufacturing was 1973. **(3 marks)**

(iii) Define the term 'tertiary industry'.

....   **(1 mark)**

Globalisation and investment

1 Study **Figure 1**.

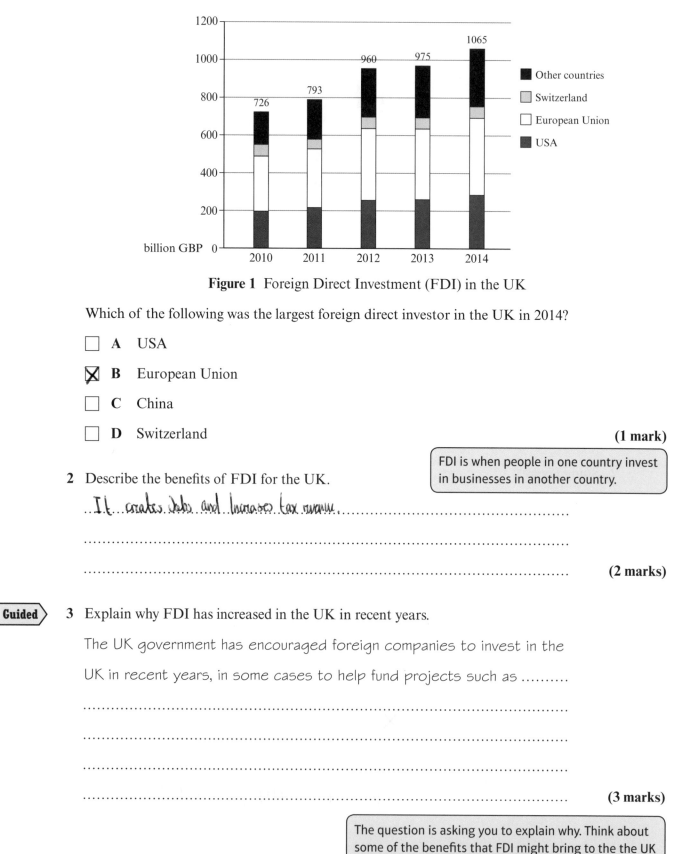

Figure 1 Foreign Direct Investment (FDI) in the UK

Which of the following was the largest foreign direct investor in the UK in 2014?

☐ **A** USA

☒ **B** European Union

☐ **C** China

☐ **D** Switzerland

(1 mark)

2 Describe the benefits of FDI for the UK.

.....It...crates..Jobs..and..Increases..tax..revenue...

....

....

(2 marks)

> FDI is when people in one country invest in businesses in another country.

> **Guided**

3 Explain why FDI has increased in the UK in recent years.

The UK government has encouraged foreign companies to invest in the

UK in recent years, in some cases to help fund projects such as ....

....

....

....

....

(3 marks)

> The question is asking you to explain why. Think about some of the benefits that FDI might bring to the the UK and explain why the government has encouraged FDI.

🌐 Case study **A UK city in context**

1 With reference to a named city in the UK, explain **one** way its location has been important to its growth and development.

....

....

.... **(3 marks)**

2 Study **Figure 1**.

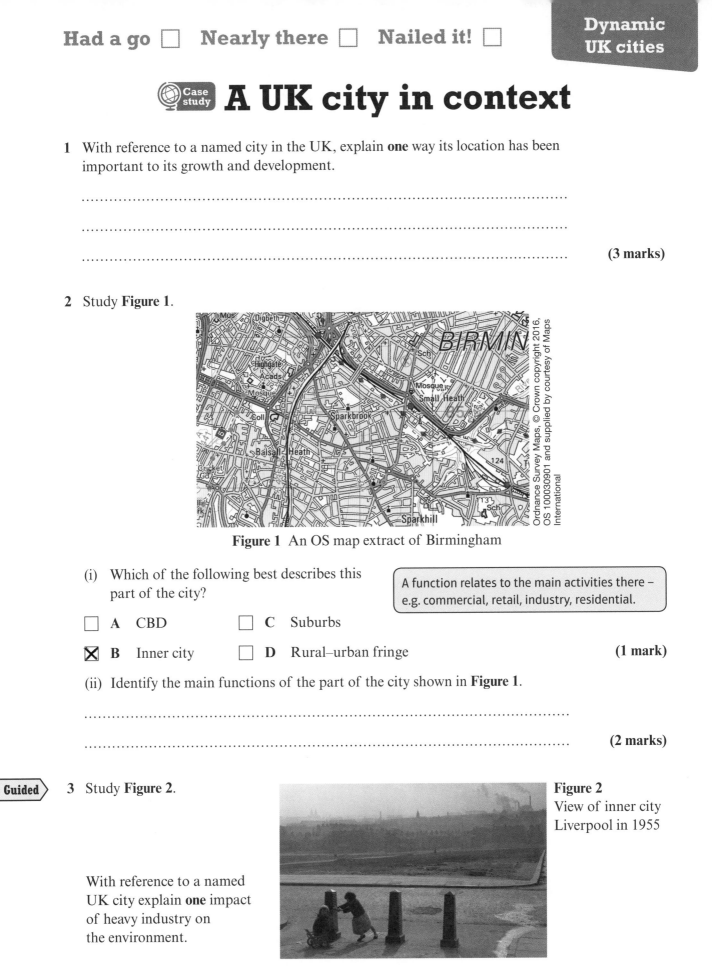

Figure 1 An OS map extract of Birmingham

Ordnance Survey Maps. © Crown copyright 2016, OS 100030901 and supplied by courtesy of Maps International

(i) Which of the following best describes this part of the city?

☐ **A** CBD   ☐ **C** Suburbs

☒ **B** Inner city   ☐ **D** Rural–urban fringe   **(1 mark)**

> A function relates to the main activities there – e.g. commercial, retail, industry, residential.

(ii) Identify the main functions of the part of the city shown in **Figure 1**.

....

.... **(2 marks)**

Guided **3** Study **Figure 2**.

Figure 2
View of inner city Liverpool in 1955

With reference to a named UK city explain **one** impact of heavy industry on the environment.

In Birmingham the metal industry polluted ....

....

.... **(3 marks)**

Case study Urban change differences

1 Study **Figure 1**.

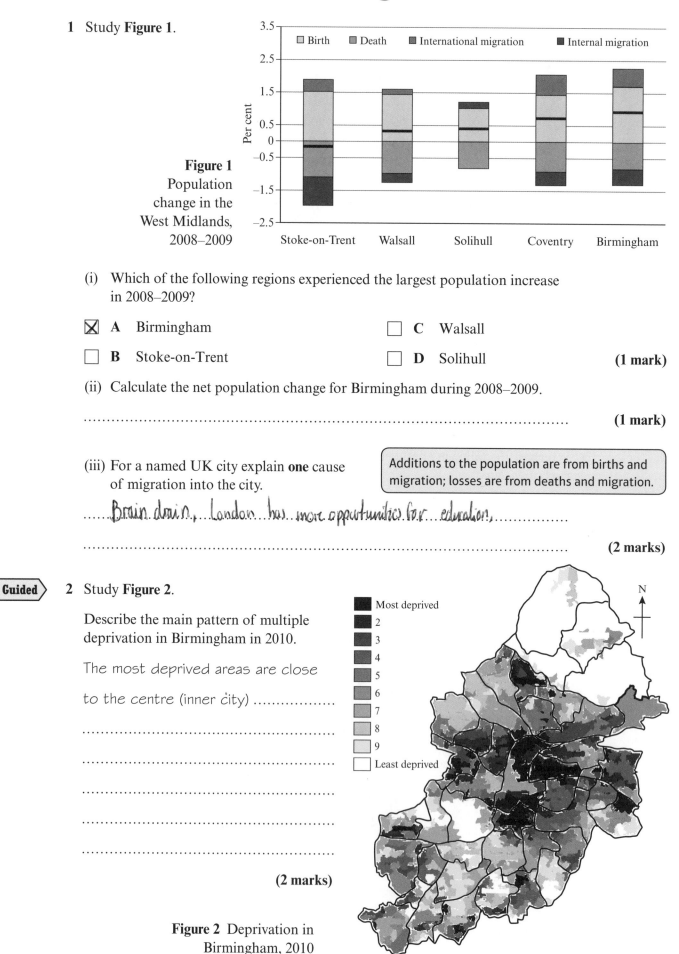

Figure 1
Population change in the West Midlands, 2008–2009

(i) Which of the following regions experienced the largest population increase in 2008–2009?

☒ **A** Birmingham   ☐ **C** Walsall

☐ **B** Stoke-on-Trent   ☐ **D** Solihull   **(1 mark)**

(ii) Calculate the net population change for Birmingham during 2008–2009.

.... **(1 mark)**

(iii) For a named UK city explain **one** cause of migration into the city.

Brain drain, London has more opportunities for education,

.... **(2 marks)**

> Additions to the population are from births and migration; losses are from deaths and migration.

> **Guided**

2 Study **Figure 2**.

Describe the main pattern of multiple deprivation in Birmingham in 2010.

The most deprived areas are close

to the centre (inner city) ....

....

....

....

....

(2 marks)

Most deprived
2
3
4
5
6
7
8
9
Least deprived

Figure 2 Deprivation in Birmingham, 2010

🌐 Case study City challenges and opportunities

1 Define the term 'de-population'.

...... the decrease of population . ... **(1 mark)**

2 Which of the following is a cause of de-population in cities?

☐ **A** Large industrial units locating in the inner city

☐ **B** TNCs locating operations in the CDB

☒ **C** Out migration to the suburbs

☐ **D** Decentralisation of retailing to the outer suburbs **(1 mark)**

3 With reference to a UK city you have studied, explain why economic decline has occurred.

> Your answer should refer to job opportunities.

....

....

....

.... **(3 marks)**

Guided **4** Define the term 'gentrification'.

Gentrification is the improvement of housing in an area that

...was previously neglected area and it causes land/house price...

...to increase. .... **(1 mark)**

5 Referring to a UK city you have studied, explain why there has been an increase in the population in certain parts of the city.

....

....

....

....

.... **(4 marks)**

> Population can increase as the result of a rising birth rate when more babies are born and/or when more people move into the city.

Case study Improving city life

1 Define the term 'rebranding'.

...The....creation of a for new name for an existing organization....................... **(1 mark)**

> **Guided**

2 Study **Figure 1**.

Figure 1 Athletes' Village accommodation blocks built
for the 2012 Olympics in East London.

For a named UK city, evaluate the sustainability of strategies to improve quality
of life.

> This answer can be split into: homes, jobs, recreation and environment.
>
> Remember, to **evaluate** is to give the advantages and disadvantages of the strategies
> and then come to a conclusion.
>
> You must refer to sustainability, which is about how the strategies meet the needs of
> people today, without comprising the ability of future generations to meet their own
> needs. For example, Figure 1 shows how parts of east London were improved by the 2012
> Olympics. If you were writing about London, you could evaluate the effects of this on the
> quality of life of those living in the area.

In the UK city I have studied, which is ....

....

....

....

....

....

....

....

.... **(8 marks)**

> Continue your answer on your own paper. You should
> aim to write approximately one side of A4 in total.

🌐 Case study **The city and rural areas**

1 Which of the following correctly defines the movement of people from urban areas to rural areas?

☐ **A** Urbanisation

☒ **B** Counter-urbanisation

☐ **C** Decentralisation

☐ **D** Suburbanisation

(1 mark)

> **Guided**

2 State **two** ways cities depend on nearby rural areas.

(i) Urban residents depend on food produced in nearby rural areas.

(ii) ....

....

(2 marks)

3 Study **Figure 1**.

Figure 1 Cartoon showing rural issues

Explain **one** environmental cost of the inter-relationship between cities and rural areas.

.There's more pollution because vehicles have to travel further.........

....

....

(2 marks)

> The **inter-relationship** is about the links between places. Take care to identify the focus of the question. In this case, it is asking just for **environmental** problems.

Case study **Rural challenges and opportunities**

1 Which of the following is the best description of rural diversification?

☐ **A** Farmers decide to grow different types of crops

☐ **B** Farmers decide to allow housing development on their land to make money

☒ **C** Farmers try new activities, other than farming, to make money ✓

☐ **D** Farmers get new jobs in factories

(1 mark)

2 Explain **one** reason why some farmers in the UK have diversified their activities.

> The command here is **explain**, so you need to give a reason.

....

....

....

(2 marks)

3 Study **Figure 1**.

Local Authority	Second homes as a percentage of all homes	Number of second homes
South Hams, Devon	9.8	4113
North Cornwall	9.6	4000
Berwick-upon-Tweed	9.5	1344
North Norfolk	9.1	4753
Penwith, Cornwall	8.5	2779
South Lakeland, Cumbria	7.2	3743
Scarborough, North Yorks	7.2	3952
Purbeck, Dorset	6.9	1480
Great Yarmouth, Norfolk	5.9	2701
West Yorkshire	5.7	985

Figure 1 Number of second homes in certain rural areas of the UK

Suggest how the changes in **Figure 1** represent challenges to rural areas.

When people from urban areas buy second homes in rural areas, it puts

pressure on housing. This can have the effect of ....

....

....

....

....

....

(4 marks)

Investigating dynamic urban areas: developing enquiry questions

TASK: Investigating how and why quality of life varies within urban areas.

Only work through this page and the two that follow if you did urban fieldwork. **If you did rural fieldwork, turn to page 80.**

Guided 1 Study **Figure 1**.

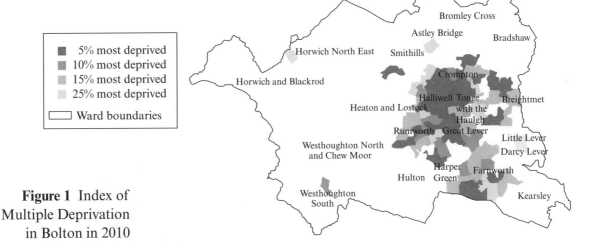

5% most deprived
10% most deprived
15% most deprived
25% most deprived
☐ Ward boundaries

Figure 1 Index of Multiple Deprivation in Bolton in 2010

(i) Suggest why the area shown in **Figure 1** might be a good place to investigate how and why quality of life varies within urban areas.

Bolton is an urban area with a range of areas of contrasting levels of

....

....

.... **(3 marks)**

(ii) State **one** enquiry question relating to the investigation task that you devised for your fieldwork.

.... **(1 mark)**

(iii) Suggest **two** contrasting wards that could be used to investigate variations in quality of life in urban areas.

....

.... **(2 marks)**

2 Explain how you selected the sites you used in your urban environment investigation.

Did you start by looking at maps, photos or data?

....

....

.... **(3 marks)**

Investigating dynamic urban areas: techniques and methods

1 Which of the following is an example of a qualitative fieldwork method?

☐ **A** Land use survey

☒ **C** Annotated field sketch

☐ **B** Pedestrian count

☐ **D** Traffic count

(1 mark)

2 Explain **one** weakness of the technique you used to investigate environmental quality.

> Environmental surveys are subject to personal viewpoints.

It was personal opinion so it's not reliable.

....

(2 marks)

Zone of urban land-use, showing some retail, residential and transport functions. At the time the photo is taken there are relatively low traffic flows and low pedestrian densities, suggesting the image is taken either away from the core CBD area, or at a time (e.g. early morning) when there are fewer people around. The residential dwellings are typical of 1960s upwards build (tower blocks) enabling high density housing in areas of higher bid-rent. Nowadays, this type of accommodation is often viewed as less desirable and is sometimes associated with a higher than average incidence of crime.

The retail outlets look to be of a similar age, perhaps 1970, with a typical mix of concrete and blockwork giving a utilitarian landscape. There is evidence of modernisation in the scene with the addition of coloured steelwork, together with more recent fascia boards.

Figure 1
Analysis of photographic evidence from an urban area

Guided > **3** Study **Figure 1**. Explain why **Figure 1** shows a good method of investigating quality of life in urban areas.

The analysis identifies the land use and the image ....

....

(2 marks)

4 Study **Figure 2**.

Indicator	Rumworth	Bradshaw	Bolton
Occupation			
Professional	10.8%	18.6%	14.9%
Unskilled	17.0%	9.0%	11.0%
Unemployed	2.7%	1.5%	2.0%
Education qualifications			
Degree level qualified	15.4%	28.0%	22.2%
No qualifications	35.6%	21.3%	26.3%

Figure 2 Census data from two wards in Bolton

Describe how the information in **Figure 2** could be used to investigate how and why quality of life varies within urban areas.

....

....

....

(3 marks)

Investigating dynamic urban areas: working with data

1 Study **Figure 1**.

Indicator	Rumworth	Bradshaw	Bolton
Occupation			
Professional	10.8%	18.6%	14.9%
Unskilled	17.0%	9.0%	11.0%
Unemployed	2.7%	1.5%	2.0%

Figure 1 Census data from two wards in Bolton

Suggest **one** way this data could be presented.

..Multiple bar charts......................................

> Try to select the most appropriate technique. What techniques did you use?

....   **(1 mark)**

2 (i) Describe **one** technique you used to present the secondary data you used in your urban study.

...............Bi   Barcharts...........................

(ii) Explain **one** weakness of the technique.

......If the data is in percentages so it's harder to......put into a comparitive bar chart,............................   **(2 marks)**

> **Guided**

3 Study **Figure 2**.

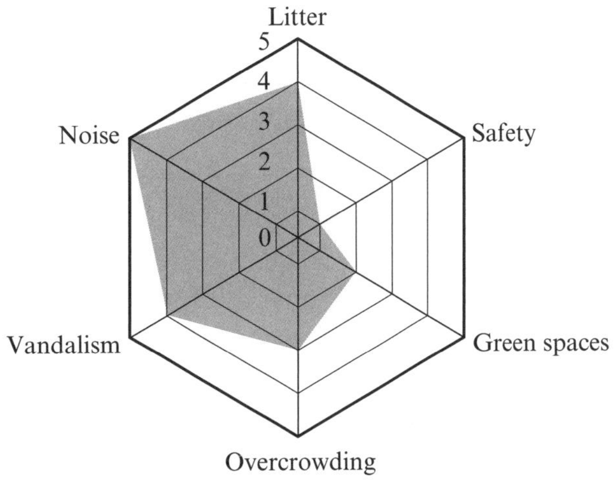

Figure 2 A graph showing environmental quality in an urban area (0 = negative, 5 = positive)

Evaluate the suitability of this data presentation technique to investigate variations in the quality of life in urban areas.

One of the advantages of the radar graph is that multiple criteria can be

shown. However, there are some problems with this type of graph ....

....

....   **(8 marks)**

> Continue your answer on your own paper. You should aim to write approximately one side of A4 in total.

Investigating changing rural settlements: developing enquiry questions

> **TASK: Investigating how and why deprivation varies within rural areas in the UK.**

> Only work through this page and the two that follow if you did rural fieldwork. **If you did urban fieldwork, turn to page 77.**

1   Study **Figure 1**.

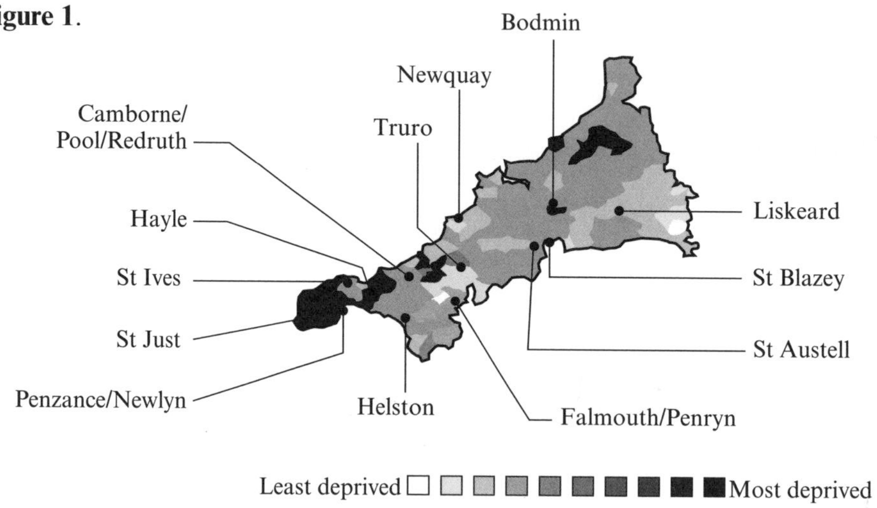

Least deprived ☐ ☐ ■ ■ ■ ■ ■ ■ ■ Most deprived

Figure 1 Areas of deprivation in Cornwall

(i)   Explain why the area shown in **Figure 1** might be a good place to investigate how and why quality of life varies within rural settlements.

....

....

....

....   **(3 marks)**

(ii) State **one** enquiry question relating to the investigation task that you devised.

> This should be about the type of things you wanted to investigate.

....

....   **(1 mark)**

(iii) Suggest **two** contrasting areas that could be used to investigate variations in quality of life in rural areas.

....   **(2 marks)**

2   Explain how you selected the sites you used in your rural environment investigation.

> Did you use any maps, photos and data to help you?

....

....

....   **(3 marks)**

Investigating changing rural settlements: techniques and methods

1 Which of the following is an example of a qualitative fieldwork method?

☐ **A** Land use survey

☐ **B** Pedestrian count

☐ **C** Annotated field sketch

☐ **D** Traffic count

(1 mark)

2 Study **Figure 1**.

Figure 1
A deprived rural village

Add annotations to **Figure 1** to describe the environmental quality in this area.   **(3 marks)**

> **Guided**

3 Study **Figure 2** (refer to the map on page 80 for the location of these areas).

Indicator	Treneere (North Penzance)	Truro	Cornwall
Occupation			
Professional	10.7%	18.8%	13.8%
Unskilled	14.0%	13.8%	12.0%
Unemployed	1.4%	1.5%	1.2%
Education qualifications			
Degree level qualified	20.1%	28.9%	25.0%
No qualifications	26.2%	22.4%	22.4%
Housing type			
Detached	17.0%	21.8%	38.2%
Terraced	38.0%	32.3%	22.6%

Figure 2 Census data for three areas in Cornwall

Describe how the information in **Figure 2** could be used to investigate how and why quality of life varies within urban areas.

The data are useful because they include information about the type of

work people do and how skilled they are, and therefore ....

....

....   **(3 marks)**

Investigating changing rural settlements: working with data

1 Study **Figure 1**.

Indicator	Treneere (North Penzance)	Truro	Cornwall
Occupation			
Professional	10.7%	18.8%	13.8%
Unskilled	14.0%	13.8%	12.0%
Unemployed	1.4%	1.5%	1.2%

Figure 1 Census data from two areas in Cornwall, Treneere and Truro

> Try to apply some of the data presentation methods you used to this question.

Suggest **one** way this data could be presented.

.... **(1 mark)**

2 (i) Describe **one** technique you used to present the secondary data you used in your rural study.

....

(ii) Explain **one** weakness of the technique.

....

.... **(2 marks)**

 3 Study **Figure 2**.

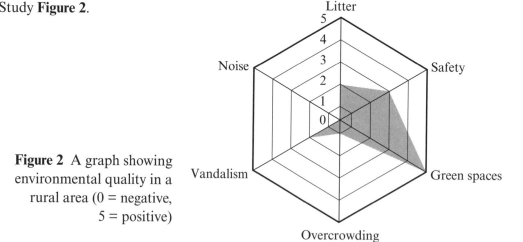

Figure 2 A graph showing environmental quality in a rural area (0 = negative, 5 = positive)

Evaluate the suitability of this data presentation technique to investigate variations in the quality of life in rural areas.

One of the advantages of the radar graph is that multiple criteria can be

shown. However, there are some problems with this type of graph

....

.... **(8 marks)**

> Continue your answer on your own paper. You should aim to write approximately one side of A4 in total.

Paper 2

1 Study **Figure 1**.

The student carried out environmental surveys at three sites in an inner city ward and three sites in a suburban ward. The student also analysed data from the 2011 census.

My findings

The highest quality environments were found in the suburbs, where there are more detached houses. There were more higher-paid professional people living in the suburbs.

There were higher levels of unemployment and numbers of people with no qualifications living in the inner city.

The inner city had higher levels of multiple deprivation than the suburbs.

Figure 1 A summary of the results from a student's investigation of how and why quality of life varies within an urban area

Evaluate the student's method and findings.

In this question you need to evaluate using a resource. You need to weigh up the advantages and disadvantages of the methods used to collect the data and the results presented in Figure 1.

For the best answer, you will need to:

- apply and deconstruct information and write a logical, balanced argument
- make judgements that are supported by evidence
- demonstrate you know and can apply a range of geographical and enquiry skills
- communicate enquiry-specific fieldwork findings
- use relevant geographical terminology consistently.

....

....

....

....

....

....

....

....

....

....

(8 marks)

Continue your answer on your own paper. You should aim to write approximately one side of A4 in total.

Distribution of major biomes

1 Define the term 'biome'.

.....An area of a aquatic ecosystem.. (1 mark)

2 Study **Figure 1**.

Figure 1 A map of the major biomes

Identify which type of biome area X is.

☐ **A** Tundra   ☒ **C** Taiga

☐ **B** Temperate forest   ☐ **D** Grassland   (1 mark)

3 Describe the global distribution of tropical rainforest.

> Area Y in Figure 1 is a tropical rainforest. When describing global distribution, state the main areas where rainforests are found and refer to latitude – where they are in relation to the equator.

.....the majority of tropical rainforests are around the equator.....................

(2 marks)

4 Study **Figure 2**.

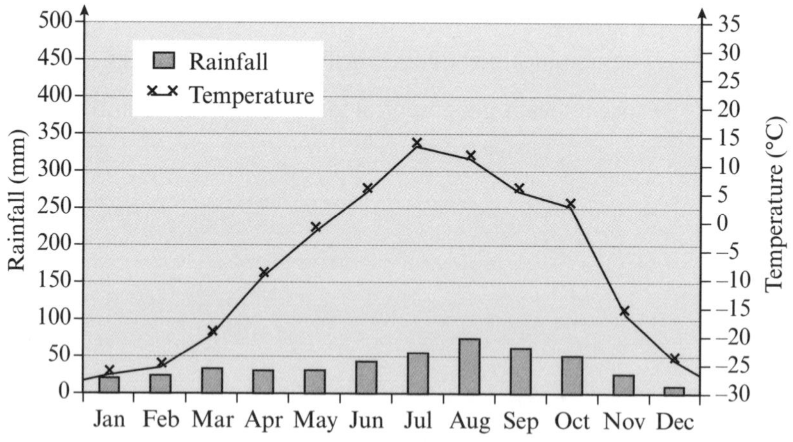

Figure 2 A climate graph of a biome

State the biome that would develop (grow) in these climatic conditions.

.... (1 mark)

Guided 5 Compare the climate shown in **Figure 2** with a tropical climate.

The climate shown in Figure 2 has a large annual

> There is a climate graph of the tropical biome on page 1.

temperature range of 42 °C, with winter temperatures falling to −27 °C

and summer temperatures reaching 15 °C. This contrasts with ....

....

.... (3 marks)

Had a go ☐   **Nearly there** ☐   **Nailed it!** ☐

Local factors

1   Study **Figure 1**.

Figure 1 Photo showing part of Tanzania in East Africa, with Mount Kilimanjaro in the background

Identify the biome shown at X in **Figure 1**.

☐   **A**   Taiga

☒   **B**   Savanna grassland

☐   **C**   Tropical rainforest

☐   **D**   Desert

(1 mark)

2   Describe the differences in the vegetation between areas X and Y in **Figure 1**.

> Use evidence from the photograph.

In area X there is a lot of vegetation, and grassland, this is contrary to Y where there is a lack of vegetation due to the altitude

(2 marks)

⟩ **Guided** ⟩   3   Give **two** local factors that explain why there is very little vegetation at area Y, shown in **Figure 1**.

The altitude at area Y is much higher. This would mean temperatures

would be lower. The presence of snow would indicate temperatures

below 0°C, which results in ....

....

....

....

(4 marks)

4   Give **one** abiotic factor that explains why there is more vegetation at area X, shown in **Figure 1**.

....

....

....

(2 marks)

Biosphere resources

1 State **two** ways the biosphere can be used commercially.

> **Commercial** means where the resources will be sold for profit, often by large companies.

....

....

....

(2 marks)

2 Study **Figure 1**.

Figure 1 The Tucurui Dam in Brazil

Identify which commercial activity is shown in **Figure 1**.

....

....

....

(2 marks)

3 Explain **one** reason why this is a suitable area to construct a dam for commercial purposes.

....

....

(2 marks)

Guided **4** Explain how the biosphere can provide for indigenous and local people.

Tropical rainforests, for example, are important in providing a range of

different timber goods (especially hardwoods) that people use for

building houses and making furniture. Tropical rainforests can also

....

....

....

....

(4 marks)

Biosphere services

1 State **two** ways in which the biosphere
provides globally important services.

Think about the essential services for life on
Earth, the air we breathe, food and water.

....

....

.... **(2 marks)**

Guided 2 Explain **one** way in which the biosphere maintains soil health.

Trees and plants provide leaf litter to the soil, which

....

....

.... **(2 marks)**

3 Explain **one** reason why the biosphere is globally important.

....

....

.... **(2 marks)**

4 Explain how changes to the biosphere affect the hydrological cycle.

....

....

....

....

....

.... **(4 marks)**

5 Explain **one** way in which the biosphere
regulates the composition of the
atmosphere.

Oxygen and Carbon Dioxide are two of the main gases of
the atmosphere. How does the biosphere affect this?

....

....

.... **(2 marks)**

Pressure on resources

1   Study **Figure 1**.

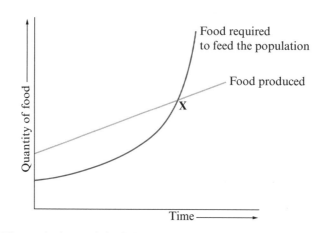

Figure 1 A model of the Malthusian population theory

Which **one** of the following best describes what happens at point X?

☐ **A**   Food produced is equal to food required to feed the population

☒ **B**   Food produced is less than food required

☐ **C**   Food required is less than food produced

☐ **D**   Food produced decreases over time   **(1 mark)**

2   Using **Figure 1**, describe how the model shows the trend between 'food production' and 'food required'.

> Make sure you use ideas from Figure 1 to develop your answer. Words and phrases such as 'rising', 'falling', 'trend', 'change', 'increase' and 'over time' may also be useful.

...this diagram shows that both the....................................

...food produced and the food required increase but the................................

...food required increases faster...

....   **(3 marks)**

3   Suggest **two** reasons why the demand for food and energy resources is increasing.

...An increasing population..

...More affluent areas...   **(2 marks)**

⟩ **Guided** ⟩   4   Explain Boserup's theory of resources and population change.

Boserup's theory is basically that population growth has a positive

impact on resource supply .....because humans will innovate.................

a new way to provide food the point at which they meet is the......

...point when a solution is found..

....   **(4 marks)**

Tropical rainforest biome

1 Which of the following is an abiotic component of an ecosystem?

☐ **A** A tree

☐ **B** An insect

☒ **C** The soil

☐ **D** People

(1 mark)

> **Guided**

2 Study **Figure 1**.

Figure 1
Part of the tropical rainforest

Identify **two** features of the rainforest shown.

(i) On the left of the picture there are leaves

with drip tips.

(ii) In the centre of the picture, ...

....

(2 marks)

> **Identify** requires you to select and name. Don't be tempted to explain the adaption.

3 Explain **one** way animals have adapted to the rainforest.

....Slaths have adapted to not grow hair an their feed so they have a.

....better grip..

(2 marks)

4 Explain how the tropical climate influences the plants and animals of the tropical rainforest.

> Remember that the tropical climate is hot and wet all year long.

....

....

....

....

(3 marks)

Taiga forest biome

1   Define the term 'biotic characteristic'.

(1 mark)

....

....

2   Study **Figure 1**.

Figure 1
Brown bears in the taiga forest

Suggest **one** way the animals shown in **Figure 1** have adapted to the climate in that area.

> The seasonal nature of the climate is an important factor in this answer.

...They have thick hair to insulate themselves in the cold.........................

.... seasons .... **(2 marks)**

> **Guided**

3   Explain **two** ways the plants of the taiga have adapted to the climate in that area.

(i)   The trees have needle-shaped leaves, which remain on the trees

throughout the year (evergreen coniferous). This means they can

photosynthesise throughout the year and intercept water all year

long.

(ii) ....

....

....

.... **(4 marks)**

> Coniferous trees, which have cones and needles that create an acidic litter, are found in the taiga forests. How will this help these trees survive the climate?

Productivity and biodiversity

1 Which of the following best defines the term 'biodiversity'?

☒ A   The variety of different living species in an area

☐ B   The rate at which new living species are produced

☐ C   A system of interlocking and interdependent food chains

☐ D   The recycling of dead organic matter and inorganic matter back into the
production of living matter
(1 mark)

> **Guided**

2 Study **Figure 1**.

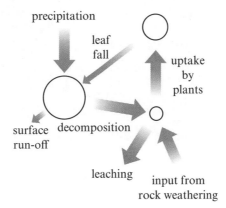

Figure 1 Nutrient cycles for the taiga ecosystem

(i)   Add labels (B, L and S) to show the biomass, litter and soil storage
components of the nutrient cycle of the taiga.
(2 marks)

(ii) Compare the nutrient cycle of the taiga with the nutrient cycle in tropical
rainforests.

The tropical rainforest has very high nutrient storage in the biomass

compared with smaller biomass storage in the taiga. The tropical

rainforest and the taiga both have low ....

....

....

.... **(3 marks)**

3 Explain why very few nutrients are stored in the plant
litter in the tropical rainforest.

> You need to give reasons that are linked
> to the climate of the tropical rainforest.

....

....

....

.... **(3 marks)**

Tropical rainforest deforestation

1   Study **Figure 1**.

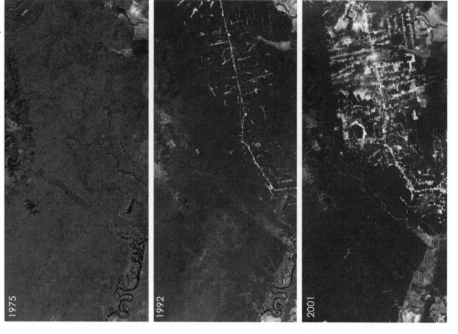

Figure 1 Satellite images of the same area of Brazilian rainforest in 1975, 1992 and 2001

Describe the changes to the area shown in **Figure 1** from 1975 to 2001.

> Try to use geographical descriptors such as north, south, east and west, and try to give approximate amounts of clearance.

....

....

....   **(2 marks)**

> **Guided**

2   Suggest **two** reasons why the area in **Figure 1** has changed since 1975.

(i)   The area may have experienced deforestation. Hardwood trees may have been removed for use as timber.

(ii)   ....

....

....

....

....   **(4 marks)**

3   Explain **one** reason why climate change causes stress to the tropical rainforest ecosystem.

....

....

....

....   **(3 marks)**

Threats to the taiga

1 Identify which of the following is most likely to be an indirect threat to the taiga.

    ☐ **A** Logging for softwoods

    ☐ **B** Deforestation for agriculture

    ☐ **C** Acid rain

    ☐ **D** Forest clearance to make way for HEP   **(1 mark)**

Guided 2 Explain how commercial development threatens the taiga.

    Removing the trees for softwood removes ....

> Commercial means large-scale activities carried out to make profit.

....

....

....

.... **(4 marks)**

3 'Climate change is the biggest threat to the taiga ecosystem.' Assess this statement.

> You need to discuss all the factors and the other threats and come to a conclusion about what is the greatest threat.

....

....

....

....

....

....

....

....

....

....

....

....

....

....

....

.... **(8 marks)**

Continue your answer on your own paper. You should aim to write approximately one side of A4 in total.

Protecting the rainforest

1   Study **Figure 1**.

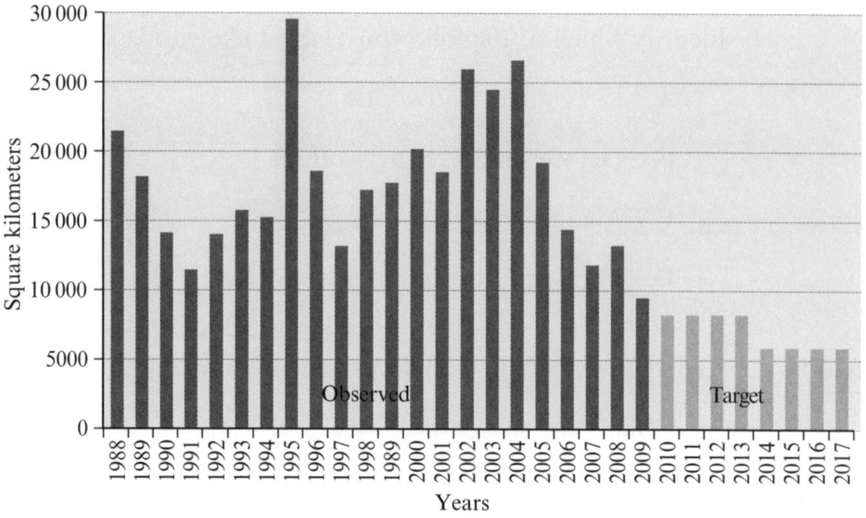

Figure 1
Deforestation in the
Brazilian Amazon.
Observed 1988–2009,
target for 2010–2017

Describe the deforestation trends in the Brazilian Amazon since 1988, shown in **Figure 1**.

> Look for when deforestation peaked, when it was increasing or decreasing, and whether it ever fluctuated. For this type of question, support your answer with figures from the graph.

....

....

....

....   **(2 marks)**

2   Suggest **one** reason deforestation rates have decreased in some areas.

....

....

....

....   **(2 marks)**

⟩ **Guided** ⟩ 3   (i)   Name **one** global action designed to protect the tropical rainforest species and areas.

....   **(1 mark)**

(ii)   Evaluate the measures that have been taken to protect the tropical rainforest species and areas.

CITES has huge global influence, with 181 countries signed up to it.

This puts pressure on countries not to hunt for endangered species

such as the rhino and therefore helps protect this species. However,

....

....

....   **(8 marks)**

> Continue your answer on your own paper. You should aim to write approximately one side of A4 in total.

Sustainable tropical rainforest management

1 Read the following newspaper article.

> The South American country of Ecuador has rich rainforest resources, which are the most biodiverse in the world. But huge reserves of oil have also been discovered beneath the Yasuni tropical rainforest National Park – worth US$7.2 billion. Instead of extracting this oil, the Ecuador government has asked the world to pay it half that amount in order to 'leave the oil in the soil'. Environmentalists have praised the Ecuador government for this proposal; after all, it is the rich countries that have damaged the world most with pollution. But others have criticised the scheme, saying it is nothing more than environmental extortion.

Identify which statement concerning the newspaper article is **true**.

> You need to check each statement carefully against the extract. There's no point rushing your answer and making a mistake.

☐ A Extracting the oil from Ecuador's Yasuni National Park would not damage the rainforest.

☐ B Ecuador has offered to save its rainforest in return for US$7.2 billion from richer countries.

☐ C The Ecuador proposal is one example of how rainforest can be conserved for the future.

☐ D There is international agreement that the Ecuador proposal is a very good idea.

(1 mark)

Guided 2 (i) Identify **one** other way tropical rainforests can be managed.

.... **(1 mark)**

(ii) Explain **one** reason why achieving sustainable forest management may be challenging.

The challenge is making sure the environment is protected, but not ......

....

....

....

.... **(3 marks)**

3 Explain why ecotourism can be a sustainable way of using the tropical rainforest.

....

....

....

.... **(2 marks)**

Protecting the taiga

1 Study **Figure 1**.

Figure 1
A measure to protect the exploitation of taiga forests.

(i) Suggest how the measure shown in **Figure 1** protects the taiga forests from being exploited.

> Develop the evidence from the photograph.

....

.... **(2 marks)**

(ii) Explain **one** of the potential limitations of the protection measure shown in **Figure 1**.

....

.... **(2 marks)**

2 Explain why taiga forests need protecting.

....

....

....

.... **(2 marks)**

Guided 3 Explain why there are conflicting views about protecting the taiga forests.

Some people believe that the taiga forests should be protected

because they are very fragile ecosystems, with many endangered

species that have adapted to the environment. Others believe that ......

....

....

....

....

.... **(4 marks)**

Energy impacts

1 State **one** example of a non-renewable energy resource.

.... **(1 mark)**

2 Identify which of the following is an example of a recyclable energy resource.

Take your time to consider all the possible answers. 'Recyclable energy' means the fuel can be re-used.

☐ **A** Wind power

☐ **B** Natural gas

☐ **C** Nuclear power

☐ **D** Hydroelectric power (HEP) **(1 mark)**

3 Study **Figure 1**.

Figure 1
Wind turbines

Suggest how the wind turbines shown in **Figure 1** could affect the landscape.

....

....

.... **(2 marks)**

Guided 4 Explain **one** negative environmental impact of mining.

Removing the vegetation on the surface of the ground and clearing

large areas of land to develop the mine. This can ....

....

....

.... **(2 marks)**

Access to energy

1   Study **Figure 1**.

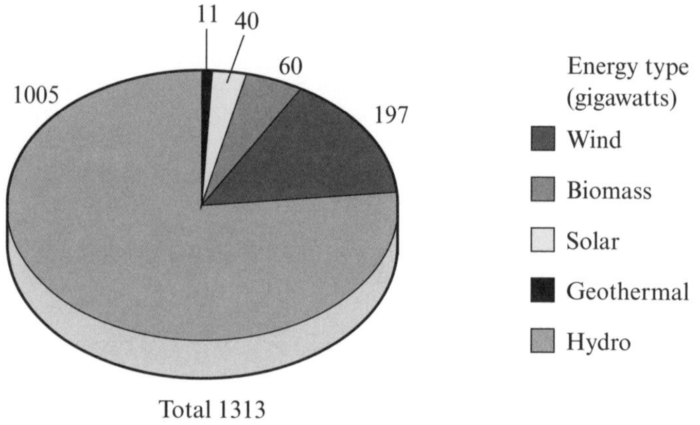

11 40
60
1005
197

Energy type
(gigawatts)

■ Wind
■ Biomass
□ Solar
■ Geothermal
■ Hydro

Figure 1 Global gigawatts of
renewable energy by type

Total 1313

What percentage of global renewable energy comes from wind power?

☐ **A** 5%   ☐ **B** 5%   ☐ **C** 25%   ☐ **D** 45%   **(1 mark)**

▷ **Guided** ▷   2   Explain why there are differences in the amount of renewable energy used
around the world.

All of these renewables depend on the natural resources available ....

....

....

....

....   **(4 marks)**

3   Study **Figure 2**.

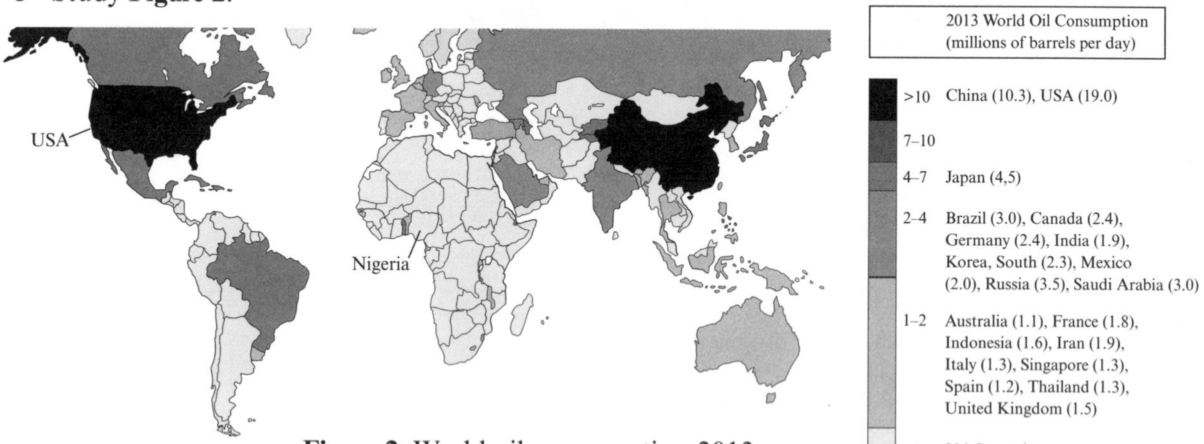

USA

Nigeria

2013 World Oil Consumption
(millions of barrels per day)

■ >10   China (10.3), USA (19.0)

■ 7–10

■ 4–7   Japan (4,5)

■ 2–4   Brazil (3.0), Canada (2.4),
Germany (2.4), India (1.9),
Korea, South (2.3), Mexico
(2.0), Russia (3.5), Saudi Arabia (3.0)

■ 1–2   Australia (1.1), France (1.8),
Indonesia (1.6), Iran (1.9),
Italy (1.3), Singapore (1.3),
Spain (1.2), Thailand (1.3),
United Kingdom (1.5)

□ <1   204 Countries

Figure 2 World oil consumption 2013

(i)   Compare the amount of oil consumed in Nigeria and the USA.

....

....

....   **(3 marks)**

(ii)   Suggest **one** reason Nigeria has lower oil consumption.

....

┌─────────────────────────────┐
│ You need to refer to levels of │
│ development in this answer.   │
└─────────────────────────────┘

....   **(2 marks)**

Global demand for oil

1   Study **Figure 1**.

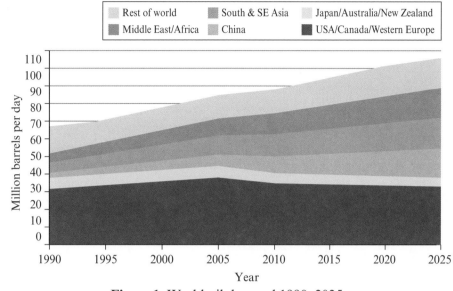

Figure 1   World oil demand 1990–2025

Using **Figure 1**, describe the changes in world oil consumption from 1990 to 2025.

> For this type of question you need to pick out the main trend and the areas with the biggest change. You need to support one of your points with data from the graph.

....

....

.... **(3 marks)**

2   Explain why there has been a significant increase in oil demand in some parts of the world.

....

....

.... **(3 marks)**

> **Guided**

3   Explain how international relations affect oil prices.

If there is continued conflict in the Middle East then this could hinder

the trade of oil from this area, causing the price to increase. If relations

between Middle Eastern countries and industrialised countries improve,

this could increase the trade between these countries and bring down

the price of oil. ....

....

....

.... **(4 marks)**

New developments

1 Which of the following is a new unconventional non-renewable energy resource?

☐ **A** Tar sand

☐ **B** Biofuel

☐ **C** Brown coal

☐ **D** Hydrogen cell   **(1 mark)**

> **Guided**

2 Explain **one** economic benefit of developing new unconventional oil and gas sources.

> The focus of the question is economic benefits. You need to discuss the positive gains from the money earned.

Extracting oil from tar sands or fracking for shale

gas will increase ....

.... **(2 marks)**

3 Evaluate the impacts of developing new conventional oil and gas resources in the ecologically sensitive area of the Arctic.

....

....

....

....

....

....

....

....

....

....

....

....

....

....

....

....

....

....

.... **(8 marks)**

> Continue your answer on your own paper. You should aim to write approximately one side of A4 in total.

Energy efficiency and conservation

1 Define the term 'energy conservation'.

.... **(1 mark)**

> **Guided**

2 Suggest **two** ways homes can be made more energy efficient.

(i) New homes can be built with thick cavity walls and filled with
     mineral wool insulation to prevent heat loss.

(ii) ....

....

.... **(4 marks)**

Figure 1 A hydrogen bus in Iceland

3 Study **Figure 1**

(i) Suggest how new technologies, such as the one shown in **Figure 1**, may help to reduce an area's carbon footprint.

> Hydrogen does not emit any CO_2 when used.

....

....

....

.... **(3 marks)**

(ii) Explain **one** problem of the new technology shown in **Figure 1**.

....

.... **(2 marks)**

Alternative energy sources

Guided 1 Study **Figure 1**.

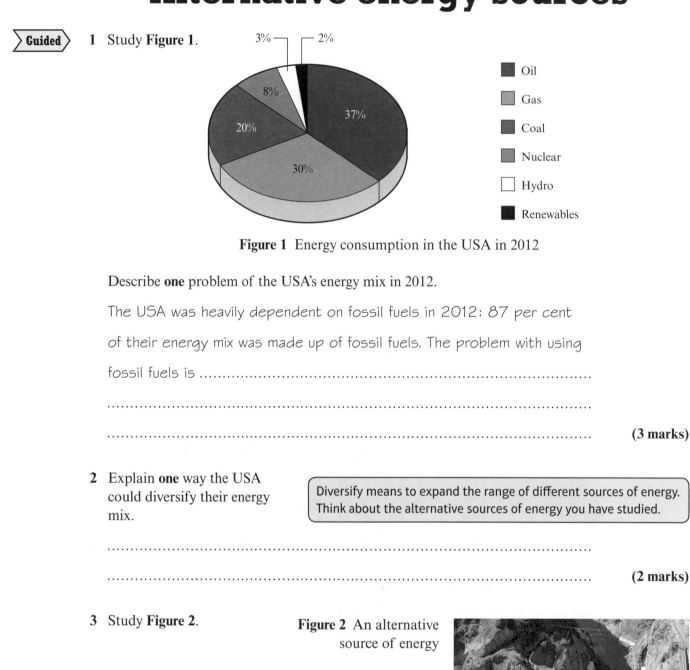

Figure 1 Energy consumption in the USA in 2012

Describe **one** problem of the USA's energy mix in 2012.

The USA was heavily dependent on fossil fuels in 2012: 87 per cent

of their energy mix was made up of fossil fuels. The problem with using

fossil fuels is ....

....

.... **(3 marks)**

2 Explain **one** way the USA could diversify their energy mix.

> Diversify means to expand the range of different sources of energy. Think about the alternative sources of energy you have studied.

....

.... **(2 marks)**

3 Study **Figure 2**.

Figure 2 An alternative source of energy

(i) Identify the type of alternative energy source shown in **Figure 2**.

.... **(1 mark)**

(ii) Explain **one** disadvantage of the energy source shown in **Figure 2**.

....

.... **(2 marks)**

Attitudes to energy

1   Assess the reason why some groups are more in favour of the development of sustainable energy resources than others.

> Half of the marks will be awarded for applying your knowledge and understanding of this issue and providing a logical, balanced and well development argument. The other half will be for using geographical skills to provide the evidence to support your points. You will also need to make an overall judgement about which of the views you think are strongest. You will need to support your judgement with evidence.
>
> Plan your answer to consider the different views, why different groups support them and what the evidence is. Think about the scientific community, TNC, governments and the public, and consider different factors: education, rising affluence and environmental concerns.

....

....

....

....

....

....

....

....

....

....

....

....

....

....

....

....

....

....

....

(8 marks)

> Continue your answer on your own paper. You should aim to write approximately one side of A4 in total.

Paper 3 (i)

1 Study all of the resources on page 120.

Using evidence, assess the local and national benefits of developing hydroelectric power in Amazonia.

> For this question you are being asked to 'assess with a resource'. You need to assess the range of benefits, both for the whole of Brazil and the local area, Amazonia. You also need to consider the disadvantages.
>
> Consider the following factors for your answer.
>
> - Nationally, how can Brazil's economy benefit from further HEP schemes?
> - Nationally, what are the advantages of HEP?
> - How might local people living in the area benefit from the construction and operation of the HEP plant?
> - How might local people be disadvantaged?
> - How might the people of Brazil be disadvantaged by the HEP scheme?
> - What are the economic and environmental costs?
>
> For the best answer, you will need to:
>
> - use the information in the resource, selecting facts and figures from the resources provided
> - write a balanced answer – in this case, addressing both the local and national benefits and disadvantages
> - support your points with evidence
> - use your wider geographical knowledge.
>
> Remember: you **must** come to a clear conclusion. There is no set answer, but you must make a reasoned judgement and support this with evidence.

....

....

....

....

....

....

....

....

....

....

....

....   **(8 marks)**

> Continue your answer on your own paper. You should aim to write approximately one side of A4 in total.

Paper 3 (ii)

1   Study the **three** options below for how Brazil should develop the rainforest.

> **Option 1:** Build the Belo Monte Dam and further large HEP schemes to produce more HEP.
>
> **Option 2:** Make the area a national park and increase the amount of energy produced from biofuels.
>
> **Option 3:** Develop smaller HEP dams and solar and wind power.

Select the option that you think would be the best **long-term** plan for the development of the Brazilian rainforest. Justify your choice.

Use information from the resources on page 120 and knowledge and understanding from the rest of your geography course of study to support your answer.

> For a strong answer you need to include the following:
>
> - **range** – refer to at least two advantages and two disadvantages (costs/benefits, good/bad, positive/negative impacts) of your chosen option
>
> - **detailed evidence** – make extended explanations using detail from the booklet, rather than vague assertions
>
> - **counter-argument** – explain why you have rejected one or more options, but also consider one of its strengths
>
> - **balance** – refer to both people and/or environment in terms of advantages and disadvantages
>
> - **synoptic** – bring in some knowledge and understanding from 'People and the biosphere' and 'Forests under threat' topics
>
> - **come to a view** – make an overall judgement that is logically linked to the evidence you have used.

Chosen option: ....
....
....
....
....
....
....
....
....
....

(12 marks + 4 marks for SPGST)

> Continue your answer on your own paper. You should aim to write between 1 and 2 sides of A4.

Atlas and map skills

1   Study **Figure 1** and **Figure 2**.

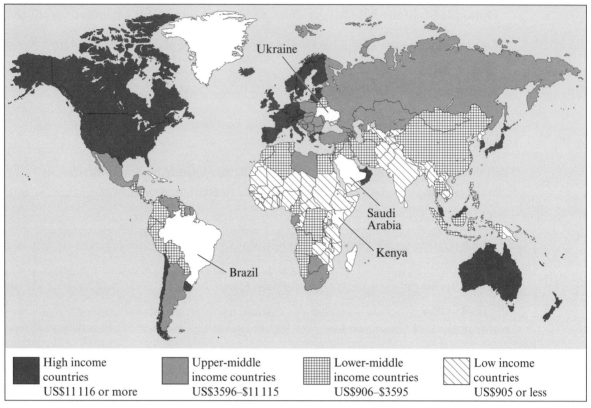

Figure 1 Global distribution of countries according to levels of income

Country	Brazil	Kenya	Saudi Arabia	Ukraine
GNI per capita (US$)	10 720	820	17 820	3120

Figure 2 GNI per capita for selected countries, 2011

(i)   Complete the map by shading in Brazil, Kenya, Saudi Arabia and Ukraine according to the key provided on the map.

> It is important to choose the right shading for each country according to the key, but your shading in doesn't need to be perfect – just clear and accurate.

(4 marks)

(ii)   Which of the following describes the type of map used in **Figure 1**?

☐   **A**   Isoline

☐   **B**   Topological

☐   **C**   Proportional symbol

☐   **D**   Choropleth   **(1 mark)**

⟩ **Guided** ⟩   2   Using evidence from the map, describe the global distribution of levels of income.

The highest income countries can be found in the continents of ....

....

....

....   **(2 marks)**

Types of map and scale

1 Study **Figure 1**.

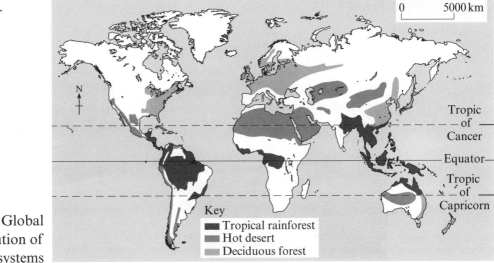

Figure 1 Global distribution of three ecosystems

Describe the distribution of hot deserts shown in **Figure 1**.

> Remember that a **describe** question does not want you to explain. For a global distribution like this, you can use latitude, hemispheres, continent names and compass directions in your description.

....

....

....   **(3 marks)**

Guided 2 Study **Figure 2**.

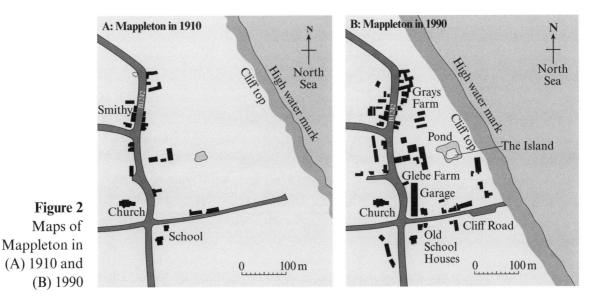

Figure 2 Maps of Mappleton in (A) 1910 and (B) 1990

(i) Calculate how many metres the cliff top receded towards Mappleton, East Yorkshire, between 1910 and 1990.

About 135 m   **(1 mark)**

(ii) Calculate how far in metres the church was from the high-water mark in 1990.

....   **(1 mark)**

Using and interpreting images

1   Study **Figure 1**.

Figure 1
Hayle Estuary
in Cornwall

(i)   Which of the following describes this type of photograph?

☐ **A**   Ground level   ☐ **C**   Vertical aerial

☐ **B**   Oblique aerial   ☐ **D**   Satellite   **(1 mark)**

(ii)   State two advantages of using a photograph like this.

Advantage 1: It is easy to compare different land uses.

Advantage 2: ....   **(2 marks)**

2   Study **Figure 2**.

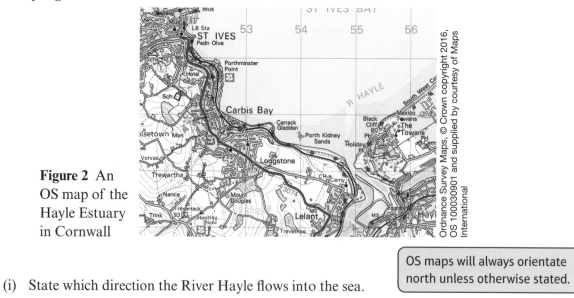

Figure 2 An
OS map of the
Hayle Estuary
in Cornwall

Ordnance Survey Maps, © Crown copyright 2016,
OS 100030901 and supplied by courtesy of Maps
International

> OS maps will always orientate
> north unless otherwise stated.

(i)   State which direction the River Hayle flows into the sea.

....   **(1 mark)**

(ii)   State **one** piece of information from the map that cannot be seen on the
photograph.

....   **(1 mark)**

Sketch maps and annotations

> **Guided**

1 The following shows a field sketch of a meander on the River Rudd.

| Fast-flowing water | Woodland | Marshland | Slow-flowing water | River cliff |

(i) Label the field sketch using the labels above. **(5 marks)**

> You can make a sketch more complex by adding annotations.

(ii) Add an annotation for 'river cliff' in the space below.

A river cliff is the steep-sided bank of the river which forms ....

....

.... **(3 marks)**

2 Draw a sketch map of the settlement of Sherburn using the OS map extract.
  Include the following details:

(i) the railway line   (iv) location of church in grid square 3142

(ii) all roads   (v) Broomside House   **(5 marks)**

(iii) the settlement of Sherburn

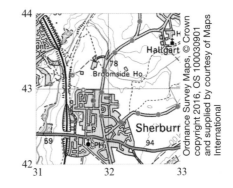

Physical and human patterns

1   Study **Figure 1**.

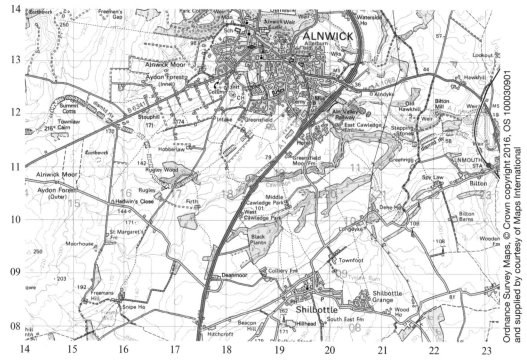

Figure 1 Map extract of Alnwick

(i)   Describe the shape of the settlement of Alnwick in the OS map extract.
Use map evidence in your answer.

Alnwick has a .... shape because ....

....

....

....

....

....   **(4 marks)**

(ii)   Suggest why Alnwick will have difficulty expanding. Use map evidence in your answer.

> Look at how the land is being used close to edges of the settlement, on the rural–urban fringe. How might this hinder the expansion of Alnwick?

....

....

....

....

....   **(4 marks)**

Land use and settlement shapes

1 Identify which of the following best describes a dispersed settlement.

☐ **A** Clustered or grouped together

☐ **B** Spread out

☐ **C** In a line along a river or road

☐ **D** On a hillside   **(1 mark)**

2 Draw the shape of:

(i) a nucleated settlement   **(1 mark)**   (ii) a linear settlement.   **(1 mark)**

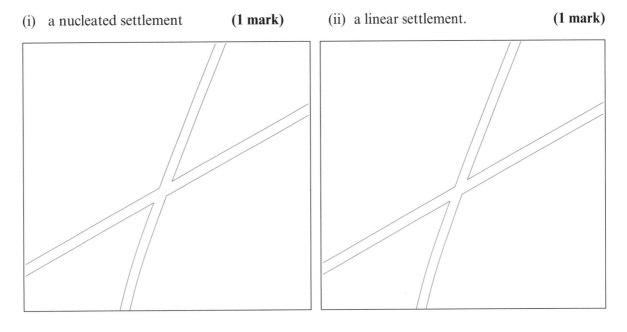

Guided 3 Use the map extract from page 113 to help you answer the following questions.

(i) Describe the shape of Sidford in grid square 1390.   **(2 marks)**

....

....

....

(ii) Describe the physical and human land uses from grid square 1290 to grid square 1390.

> You must include specific references to the map for full marks!

Use map evidence in your answer.

The land use is very rural. Brook Farm is located to the ....

....

....

....

....

....

....   **(3 marks)**

Had a go ☐ Nearly there ☐ Nailed it! ☐

Human activity and OS maps

> Use the map extract of Alnwick from page 110 to help you answer the following questions.

1 State **one** piece of map evidence in grid square 1912 that would be an important facility for local residents.

.... **(1 mark)**

2 Suggest what form of public transport can be used in grid square 1813.

> Remember that public transport would include trains, buses and trams.

....

....

....

.... **(2 marks)**

3 Shilbottle is a village to the SE of Alnwick. Identify two pieces of map evidence which prove it's a village.

> Villages will have three common 'services'. Some of these are closing down!

....

....

....

.... **(2 marks)**

Guided 4 Shilbottle is a rural settlement. Using map evidence, justify this statement.

Shilbottle is surrounded by a rural landscape. The many farms around

the area help prove this, for example, South East Farm located to the

south east of Shilbottle.

The map shows ....

....

....

....

....

.... **(4 marks)**

Map symbols and direction

Guided 1 (i) Identify what the following symbols mean.   **(2 marks)**

P [   ]

⚓🌳 [   ]

PH [   ]

⬤ [   ]

If you are not sure, look at a key on an OS map (1: 50 000)

(ii) Draw the OS symbols for the following:

Golf course 🚩

Church with a tower [   ]

Spot height [   ]

Nature Reserve [   ]

Clubhouse [   ]

Coniferous wood [   ]

(3 marks)

2 Study the map extract below.

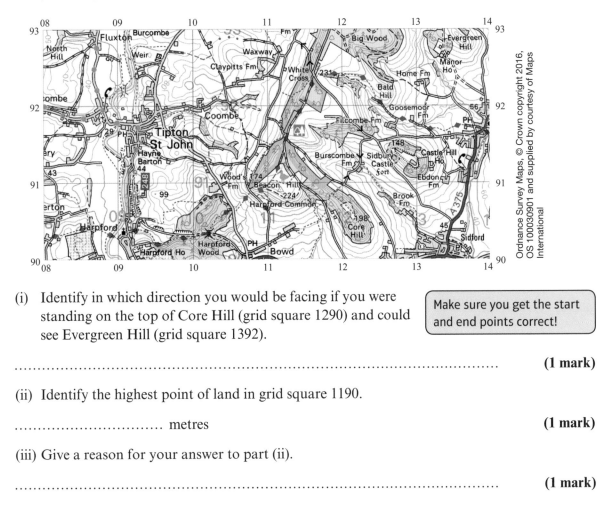

Ordnance Survey Maps, © Crown copyright 2016, OS 100030901 and supplied by courtesy of Maps International

(i) Identify in which direction you would be facing if you were standing on the top of Core Hill (grid square 1290) and could see Evergreen Hill (grid square 1392).

Make sure you get the start and end points correct!

.... **(1 mark)**

(ii) Identify the highest point of land in grid square 1190.

.... metres   **(1 mark)**

(iii) Give a reason for your answer to part (ii).

.... **(1 mark)**

Grid references and distances

> **Guided**

1 Use the map extract from page 113 to help you answer the following question.

   (i) Identify what type of woodland is shown in grid square 1292.

     D _ _ _d_ _ _ s   **(1 mark)**

   (ii) Give the 6-figure grid reference for the nature reserve in the centre of the extract.

     ....   **(1 mark)**

   (iii) Calculate to the nearest $\frac{1}{2}$ km the straight line distance from the public house in grid square 1391 to the public house in grid square 0991.

> Read the question carefully!

     .... km   **(1 mark)**

   (iv) Identify the name of the farm in grid square 1092.

     ☐ **A**   Home Farm   ☐ **C**   Brook Farm

     ☐ **B**   Claypitts Farm   ☐ **D**   Goosemoor Farm   **(1 mark)**

2 There are two bridges located on the River Otter (from grid square 0990 to 0992).

Identify the 6-figure grid reference of the most northerly bridge.

     ☐ **A**   099921   ☐ **C**   091923

     ☐ **B**   095926   ☐ **D**   095925   **(1 mark)**

3 (i) Calculate to the nearest $\frac{1}{2}$ km, the total winding distance of the River Otter.

> Remember: a piece of string will help with winding distances!

     .... km   **(1 mark)**

   (ii) Identify the symbol found at grid square 137912.

     ....   **(1 mark)**

   (iii) Identify the 6-figure grid reference for the highest point of land in grid square 1190.

     ....   **(1 mark)**

   (iv) Identify the 6-figure grid reference of the telephone which appears to be in grid square 0892.

> Make sure you look carefully. This symbol has a label line!

     ....   **(1 mark)**

Cross sections and relief

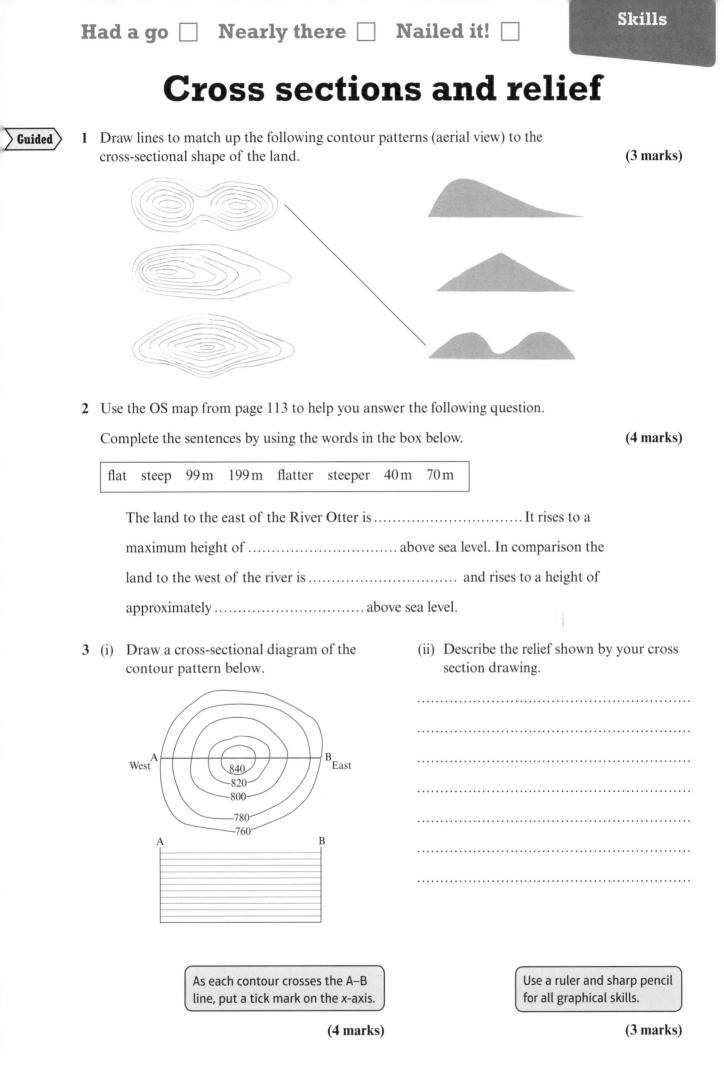

Guided 1 Draw lines to match up the following contour patterns (aerial view) to the
cross-sectional shape of the land. **(3 marks)**

2 Use the OS map from page 113 to help you answer the following question.

Complete the sentences by using the words in the box below. **(4 marks)**

flat   steep   99 m   199 m   flatter   steeper   40 m   70 m

The land to the east of the River Otter is .... It rises to a

maximum height of .... above sea level. In comparison the

land to the west of the river is .... and rises to a height of

approximately .... above sea level.

3 (i) Draw a cross-sectional diagram of the
contour pattern below.

(ii) Describe the relief shown by your cross
section drawing.

....

....

....

....

....

....

....

As each contour crosses the A–B
line, put a tick mark on the *x*-axis.

Use a ruler and sharp pencil
for all graphical skills.

(4 marks)

(3 marks)

115

Graphical skills 1

Study this graph, which shows the growth in bicycle use observed (by traffic count) in New York City.

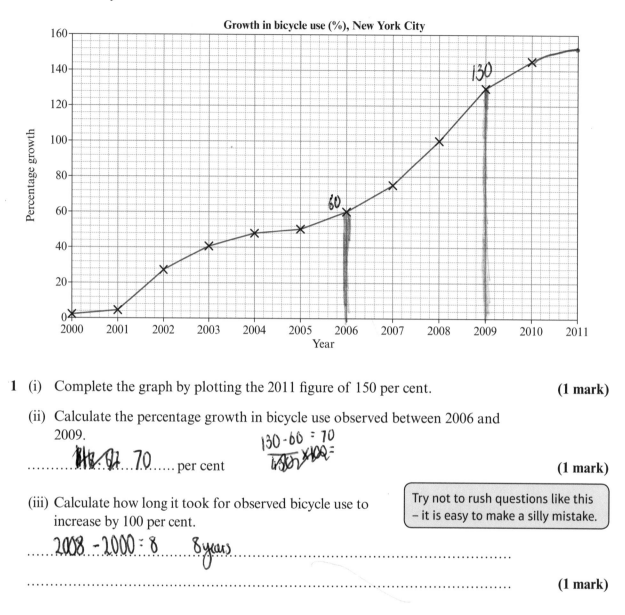

Growth in bicycle use (%), New York City

1 (i) Complete the graph by plotting the 2011 figure of 150 per cent.   **(1 mark)**

(ii) Calculate the percentage growth in bicycle use observed between 2006 and 2009.

..........70......... per cent

$130 - 60 = 70$

$\frac{70}{60} \times 100 =$ **(1 mark)**

(iii) Calculate how long it took for observed bicycle use to increase by 100 per cent.

.....2008 - 2000 = 8   8 years ....

....

(1 mark)

> Try not to rush questions like this – it is easy to make a silly mistake.

116

Graphical skills 2

Guided   **1** The diagram below shows the results of a 3-minute traffic survey completed by a GCSE class.

> Make sure that you are accurate! Use the guide lines on the diagram to help.

(i) Complete the diagram using the data in the table.

Lorry	6
Van	4

(2 marks)

(ii) Calculate how many vehicles there were in total.

.... 30 vehicles .... **(1 mark)**

2 The graph below shows the population structure of a country.

(i) State what this type of graph is called.

.... Population pyramid .... **(1 mark)**

(ii) Describe the population structure.   > Use evidence from the graph to support your answer.

there is a high birth rate, however there is a low life expectancy. because there is a large base and a skinny top.

....

....

(3 marks)

3 One student decides to present the traffic data in question 1 as a line graph. Suggest why this graphical technique is inappropriate for the data collected.

....

....

....

....

(3 marks)

Had a go ☐ Nearly there ☐ Nailed it! ☐

Numbers and statistics 1

1 Study **Figure 1**.

	2010	2011	2012	2013	2014	2015
India	1 230 984 504	1 210 193 422	1 263 589 839	1 279 498 874	1 295 291 543	1 311 050 521
World	6 929 725 043					7 349 472 099

Figure 1 Population data, 2010 to 2015

(i) Calculate the percentage of the world's population that lived in India in 2015.

Divide India's population value by the world's population value, then multiply the answer by 100.

$\frac{1311050521}{7349472099} \times 100 = 17.84\%$

(1 mark)

(ii) Calculate India's population percentage increase between 2010 and 2015.

$\frac{1311050521 - 1230984504}{1230984504} \times 100 = 6.5\%$

(2 marks)

(iii) **Figure 1** shows India's population change from 2010 to 2015. Suggest a type of graph that could best represent the data in Figure 1.

....

....

(2 marks)

Guided

2 Study **Figure 2**.

Total population:	1 210 193 422
Rural:	833 087 662
Urban:	377 105 760

Figure 2 Population by rural/urban residence in India, 2011

Calculate the ratio of rural to urban population in 2011.

833 087 662 ÷ 377 105 760 = ... This means for every 1 person living in an urban area in India ....

....

....

(2 marks)

Numbers and statistics 2

1 Study **Figure 1**.

Water year	2004–2005	2005–2006	2006–2007	2007–2008	2008–2009	2009–2010	2010–2011
Peak flow m³/s	17.7	10.7	22.9	18.6	23.2	14.8	22.6

Figure 1 Peak flow data for the River Douglas at Wigan

> **Guided**

(i) Calculate the mean peak flow over the period shown in **Figure 1**.

....

.... **(2 marks)**

(ii) Calculate the median value.

> The median is the middle value. First write the values in order from smallest to largest.

....

....

....

.... **(2 marks)**

(iii) Calculate the interquartile range for the data shown in **Figure 1**.

> The interquartile range is the difference between the lower and upper quartile values.

....

....

....

.... **(2 marks)**

(iv) Suggest how the interquartile range could be used.

The interquartile range is a measure of dispersion around the average.

It omits the very extreme values. The discharge data could be used ......

....

....

....

.... **(2 marks)**

Paper 3 (iii): resources

People and the biosphere

> Use the resources on this page to answer the question on page 104.

The issue

Brazil is an emerging country which has experienced rapid economic growth in recent years. A reliable, secure and sustainable energy supply will be crucial to Brazil's future development. Brazil has a range of energy resources:

- it produces bioelectricity from sugarcane crops
- large deposits of oil and gas have recently been discovered offshore
- there is large potential for wind and solar power
- there is potential for shale gas
- it is a large producer of HEP.

Choices now need to be made about how to expand its electricity supply.

Background to the issue

- Belo Monte, on the Xingu River, is the largest dam currently under construction in the world.
- Its final expected capacity will make it the world's third largest HEP producer.
- It will cost US$14.4 billion.
- 1 500 km² of Brazilian forest will be destroyed.
- Over 20 000 people will be forced to move away from the area.
- The Xingu River basin is home to 25 000 indigenous people from 40 ethnic groups.
- Several rare and endangered fish species are threatened by the dam.

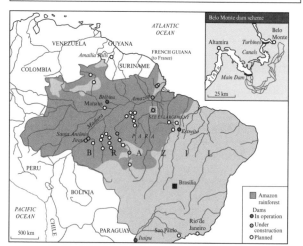

Figure 1 A map of Brazil

Figure 2 An aerial view of the Bacaja indigenous tribe settlement, on the Bacaja river banks, which is a tributary of the Xingu River

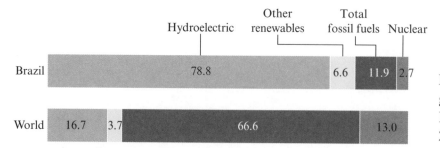

Figure 3 Electricity generated by type in Brazil and worldwide, 2010, per cent

Brazil	Hydroelectric	Other renewables	Total fossil fuels	Nuclear
Brazil	78.8	6.6	11.9	2.7
World	16.7 / 3.7		66.6	13.0

Brazilian government	Hydroelectric power is clean, green energy and we have the potential to produce high amounts of HEP, which will benefit many millions of people in Brazil.
Celebrity conservationist	The Amazon rainforest is a global resource and should be protected for future generations. I am against HEP development.
A migrant worker	I moved from the overcrowded city for a house and a well-paid construction job. I am in favour of HEP development.
An indigenous tribal person	Generations of my family have lived off the river and we may lose the traditions of our tribe. I am against HEP development.

Figure 4 Conflicting views about the development

Answers

Where an example answer is given, this is not necessarily the only correct response. In most cases there is a range of responses that can gain full marks.

Hazardous Earth

1 Global circulation

1 **B** Hot wet tropical
2 1.5 °C
3 15 °C
4 Tresco and Newfoundland are located at similar latitudes but their climates are very different. Tresco is much <u>milder than Newfoundland. Temperatures are 15 °C higher during the day. The milder temperatures mean Tresco has 0 days of snow, compared with 18 days in Newfoundland.</u>

2 Natural climate change

1 **B** Average global temperature was higher in 2010 than in 1860.
2 One natural cause is that the amount of heat the Earth receives from the sun varies because of changes to the orbital cycle, which could trigger ice ages. Another natural cause is that <u>volcanic eruptions can cause cooling of the Earth. This is because volcanoes can pump large volumes of ash and dust particles into the upper atmosphere, which can block out the sunlight and cause temperatures to drop.</u> *Other answers are possible, e.g. sunspots.*
3 The painting was created in 1565 and shows a cold climate, as there is snow lying on the ground and frozen water. This is a historical source that geographers can use to show how the climate has changed over the years.

3 Humans and climate change

1 Farming intensification has meant that large amounts of methane have been released from cattle and rice paddies. Methane traps outgoing infrared radiation in the atmosphere causing the climate to warm up. Nitrous oxides from jet engines trap infrared heat in the atmosphere causing temperatures to rise. *Other answers are possible, e.g. deforestation.*
2 Recent extreme weather events, such as <u>excessive rainfall in Bangladesh and the resulting flooding, are seen as evidence that the climate is changing.</u>
3 There is some uncertainty in natural events. There could be an increase in volcanic activity or sunspot activity. It is unclear how the oceans will respond to these. There may be a change in human activity, a switch to renewables and clean energy and transport that may mean fewer greenhouse gases in the atmosphere, which could result in the lower prediction of the sea level rise.

4 Tropical cyclones

1 They occur in the tropics between 5° and 30° latitude. They affect places such as the south-east coast of North America, coastal areas of south-east Asia and north-eastern Australia.
2 The map shows the tropical cyclones being largely restricted to the tropics. There are three main reasons for this location. First, these storms are powered by warm ocean temperatures – the seawater needs to be above 26.5 °C and these temperatures are only found in late summer and autumn in the tropics.

Second, <u>high temperatures cause warm moist air to rise over the oceans, creating intense low pressure and causing large towering cumulonimbus clouds to form and rotate because of the Coriolis force. The weather system moves landward to the cooler coastal areas.</u>
3 **C** Australia

5 Tropical cyclone intensity

1 **C** The Saffir–Simpson scale
2 Using the scale, Typhoon Haiyan was classified as a category 5 cyclone, where the damage would be <u>catastrophic destruction of property up to 5 m above sea level. A typhoon of this magnitude would mean that mass evacuation of people away from the affected area would be necessary.</u>
3 Choose one of the following: water temperatures above 30 °C; low wind shear; high humidity.
4 Typhoon Haiyan had much higher wind speeds, a taller storm surge, lower pressure and many more people were killed.

6 Tropical cyclone hazards and impacts

1 Tropical cyclones can bring extreme weather conditions such as changes in atmospheric pressure, <u>strong winds of over 100 km per hour, heavy rainfall, flooding, landslides, storm surges up to 12 m high resulting in coastal flooding.</u>
2 Low-lying coastlines are vulnerable to <u>storm surge flooding,</u> caused by <u>low air pressure within cyclones.</u>
3 People can drown in storm surges at the coast. Homes can be destroyed, so people become homeless; people may also lose their jobs, crops or livelihoods..
4 Developing and emerging countries may be more vulnerable to the impact of tropical cyclones because they may not have enough defences or effective response systems to protect people and property from the damage caused by tropical cyclones.

7 Dealing with tropical cyclones

1 **B** A barrier that will keep back storm surges
2 **A** Satellites
3 Example: The Australian government issues evacuation instruction leaflets to prepare people for a cyclone.
4 Developed countries have more money and resources to provide <u>defences and technology to forecast cyclones, and to develop evacuation procedures that protect people and property against cyclone damage.</u>

8 Tropical cyclones

1 Your answer will depend on the example chosen.
Protection method:
   - Build storm surge defences.
   - Invest in more satellite tracking technology.
   - Educate the people living in the hazard risk area about evacuation measures.

Pros and cons of the methods:
   - Flood barriers will protect the area behind them from storm surges. However, if the storm surge is greater in height than the barrier, the waves will overtop the barrier. Flood barriers are very expensive.

- Satellites will enable better forecasting of tropical cyclones but authorities will need to act on this information and inform the public to evacuate the area. Satellites cannot stop the tropical cyclones causing damage.
- Educating people can be a relatively inexpensive way of saving lives. If people board up their homes and leave the area for a safe evacuation centre, then fewer lives may be lost and there may be fewer injuries. Not everyone may be able to or want to leave their homes. Poorer countries may not have enough evacuation centres.

9 Tectonics

1 **D** Core
2 The inner core is solid, whereas <u>the outer core and lower mantle are liquid and the upper mantle is molten.</u> The inner core is hotter – over <u>5000 °C; the mantle is just 3000 °C.</u>
3 Heat from the radioactive decay of the core heats the molten rock in the mantle. Hot rock from the core rises to the Earth's surface and cools, which then causes it to drop back down towards the core. This causes a circular current, which moves the tectonic plates above.

10 Plate boundaries and hotspots

1 **A** Where the Eurasian Plate meets the North American Plate
2 Convergent plate boundary
3 As the oceanic plate subducts beneath the continental plate, this can cause <u>earthquakes, which, if they occur out at sea, can trigger tsunamis. Violent, explosive volcanoes can also occur at this type of plate boundary.</u>

11 Tectonic hazards

1 **A** Composite cone volcano
2 Plates at convergent boundaries move together. The more dense oceanic crust is subducted under the continental plate into the mantle. The crust melts and forms a magma chamber. Under pressure, the magma rises back up through the crust to erupt on to the Earth's surface as a volcanic eruption.
3 When there is subduction of the oceanic plate under the continental plate, this causes <u>friction. If the plates stick and pressure builds up, this can be released suddenly as an earthquake.</u>

12 Impacts of earthquakes

1 Choose one of the following: tsunami wave triggered by the earthquake at sea; fires started; power shortages.
2 **C** Buildings collapsing
3 (i) US$177 billion (ii) Earthquakes in some parts of the world often have a bigger impact than others because in poorer countries <u>buildings are not built to be earthquake-proof</u> OR <u>responses are not as quick and extensive as in richer parts of the world.</u> Also the location affects the impact because <u>in built-up areas there are more people at risk.</u>

13 Impacts of volcanoes

1 One impact is that ash clouds from volcanic eruptions can stop aircraft from flying, disrupting transport. Another impact is that people cannot <u>get to work, resulting in economic losses.</u> *Could also mention health issues.*
2 **A** Climate cools as volcanic ash and dust are emitted into the atmosphere
3 Here are some points you could include in your answer.
   - Places close to plate boundaries will suffer more from volcanoes, earthquakes and tsunamis.
   - More densely populated, urbanised areas are likely to suffer more damage, as there are a greater number of buildings that can collapse in the event of an earthquake, which can injure and kill people.
   - Level of development will be an important factor. Wealthier, more developed countries can afford the technology to construct buildings that can withstand the force of earthquakes (earthquake-proof) and to educate

and communicate with the people living in the area about what to do in the event of a volcanic eruption or earthquake.

14 Managing earthquake hazards

1 The measures taken to support people during and immediately after a disaster or emergency, such as an earthquake.
2 Any two from preparation, prediction and protection measures. Making buildings earthquake-proof using rubber foundations or cross-bracing. Preparing people for what to do in the event of an earthquake (take cover under a table, turn off the gas, have an emergency kit). Areas can use satellites and tiltmeters to monitor plate movements and give advice and advanced warnings of hazards.
3 Name the event. Efforts include fire and rescue services rescuing people trapped, evacuation, providing medical assistance to injured people, emergency shelters.
4 One example is putting automatic shutters on windows. These can prevent <u>glass being dropped onto people in the street below.</u> Other examples could include: fixing furniture to walls and floors to prevent them tipping over in earthquakes; counterweights or other earthquake-proofing methods help maintain the building's stability, therefore preventing it from toppling over and causing damage to transport, homes, etc. *Answers could also refer to: steel frames, flexible outer panels on building, cross-bracing, fire-resistant building materials, deep foundations, base isolators to absorb earth tremors.*

15 Managing volcano hazards

1 **C** A long-term planning measure to prepare for future eruptions
2 During a volcanic eruption, rescue and evacuation centres can be opened away from the hazard, providing <u>people with shelter, food and first aid.</u>
3 Developed countries like the USA have more money to invest in technical equipment such as tiltmeters and Global Positioning Systems to set up volcano observatories. They can also pay volcanologists to observe and make predictions, and they can set up evacuation zones. Developing countries, however, such as Bangladesh, may lack the money to invest in high tech monitoring equipment and IT communications.

Development dynamics

16 What is development?

1 **C** The total value of goods and services produced by a country in a year, divided by the population
2 The more corrupt the government, the less developed a country may be. The government may not spend money or make decisions equally. They may concentrate government funds in areas they favour and that benefit them directly.
3 GDP is a measure of how wealthy a country is: more wealth means <u>more economically developed. The wealthier the country is the more money there is available for things that make a good quality of life, e.g. good healthcare, education and housing.</u>

17 Development differences

1 **A** The number of women per 1000 who die while pregnant or after childbirth
2 The number of deaths per thousand people per year in a country.
3 (i) The wealthier a country with a higher GDP per capita, the lower the fertility rate. Italy <u>is the wealthiest country with a GDP per capita of US$35 800 and the lowest fertility rate of 1.43.</u> (ii) This may be due to wealthier nations like Italy having better health care for infants so most survive. Therefore families don't need to have lots of children to enable the survival of some. It could be because women in developed countries are more educated and choose to have a career and enjoy material things; they delay having children until later and have fewer children.

4 Choose two differences from the following: developing countries have much higher birth rates; their population pyramid has a wider base with higher numbers of children than developed countries; developed countries have higher numbers of elderly people, and a broader apex in their population pyramid.

18 Theories of development

1 Rostow's theory shows that, to develop, countries have to meet preconditions, such as having educated people to do the work. Once countries have educated people and start saving and investing money, they develop quickly, as more people are able to do skilled work, which generates more wealth. However, this theory doesn't take account of problems when economies slow down or go backwards.

2 Frank's dependency theory states that the low levels of development in poorer countries (periphery) result from the control of the world economy by rich countries (core). In Frank's theory, low-value raw materials are traded between the periphery and core. The periphery gains some wealth and development, but the core processes these into high-value products and becomes wealthy.

3 Development theories often assume that population changes are induced by industrial changes and increased wealth, but don't take account of the role of social change in determining birth rates, e.g. the education of women. The theories may not explain the early fertility declines in much of Asia in the second half of the 20th century, or the delays in fertility decline in parts of the Middle East.

4 Frank's dependency theory was written in the 1930s and does not take into account the role that modern technology has played in the development of some countries. It does not take into account other factors, such as conflict and natural disasters.

19 Types of development

1 C Small-scale projects run by local people

2 Choose any two from: large-scale projects; often funded by governments and large companies; decisions often made by governments and large companies.

3 (i) Top-down schemes are large-scale and aimed at helping large numbers of people, or even the whole country, whereas bottom-up schemes are much smaller scale and aimed at specific, often rural communities. (ii) Bottom-up schemes are more affordable, using appropriate technology that local people can install and maintain, whereas top-down schemes are usually expensive and use high technology and skilled workers.

4 Answer depends on the group of people chosen. Examples include: consumers in the developed world benefit from cheaper goods from developing countries; people who own local businesses in developing countries where TNCs outsource can gain business through contracting work from TNCs.

20 Approaches to development

1 Choose any two from the following: large-scale HEP schemes are expensive to build and can cause a country to get into debt; projects like this involve heavy engineering and technical knowledge to build and maintain – high ongoing cost; HEP has a negative impact on the environment as it involves flooding large areas, destroying wildlife habitats and agricultural land.

2 Intermediate technology: often small-scale technology that the local community can use without too much training or high costs.

3 Intermediate technology is affordable for small rural communities. It makes use of local resources and the skills of local people, so it has a low impact on the environment.

4 Women in developing countries can get better-paid, skilled jobs in TNC factories instead of in agriculture, so improving their status and increasing family income. People can benefit from improvements in the infrastructure. Suppliers gain new markets for their goods.

21 Location and context

1 Your answer will depend on the country chosen. For example, India is a large country in Asia, with the Himalaya mountains to the north. India has an extensive coastline, with the Bay of Bengal along its eastern coastline and the Arabian Sea in the west. India's neighbours are Pakistan, Bangladesh and Myanmar to the north-west, and Nepal to the north.

2 Your answer will depend on the country chosen. India imports oil and other raw materials from China and Saudi Arabia and exports manufactured goods, especially chemicals, mainly to developed countries such as the USA and the Middle East. There is very little trade with its neighbouring countries as these are less developed. There are political tensions between India and its neighbour Pakistan over the disputed area of Kashmir and HEP projects in India could reduce irrigation for farming areas in Pakistan.

3 Your answer will depend on the country chosen. Your answer should relate to how connected your country is to other parts of the world through trade, transport and TNCs.

4 Your answer will depend on the country chosen. For example, India has a variety of environments, including mountainous terrain to the north, deserts in the west, rolling plains along the Ganges, and 7000 km of coastline. The environment has attracted tourists to the country and, as a result, has created tourist jobs and improved the economy. The coastline has enabled ports to grow in areas such as Mumbai in the west of the country. Industries have developed here, creating wealth in the area. Over 12 different languages are spoken in India, and the majority of people in India – 79.8 per cent – are Hindu.

22 Globalisation and change

1 (i) Two indicators that show India has recently experienced economic growth are its GDP per capita of $5800 and the exports being mainly manufactured goods such as iron and steel. (ii) India still has a high percentage of people working in primary industry; this indicates that it is not yet a fully developed economy. (Is this true of your case study?)

2 The high percentage (almost half) of people still employed in primary industry – mainly low-paid, low-skilled work such as farming.

23 Economic development

1 (i) 2010 (ii) US$70 per dry tonne

2 When prices are increasing this makes the cost of raw materials higher for businesses, which makes the finished goods more expensive or reduces their profits. When prices fall raw materials are cheaper, which cuts costs and increases profit (or could be because of low demand therefore fewer sales and less profit).

3 TNC factories may cause air and water pollution in emerging countries. Transporting goods and people by air, road and sea contributes to increased emissions of CO_2 and other air pollutants. Using new technologies can be more efficient and reduces the need for people to travel.

4 Your answer will depend on the country chosen. For example, in the case of India, there are wide variations of GDP. Goa has a GDP per capita of 192 652 Indian Rupees, whereas in the north-east of the country, in the state of Bihar, GDP per capita is below 40 000 Indian Rupees. That is a difference of over 150 000 Indian Rupees.

24 International relationships

1 C Western Europe

2 D 500 billion dollars

3 Trade allows emerging countries to import raw materials and energy resources such as iron ores and oil to use in the new industries, and to gain wealth from export sales.

4 Your answer will depend on the country chosen. Your chosen country may have been supported by a loan from the World Bank, the government may have been given financial advice from the IMF or it may have been able to trade freely with certain countries under World Trade Organization rules.

25 Costs and benefits

1 Your answer will depend on the emerging country chosen. Here are some points you could include in your answer.
- TNCs have located their operations in many emerging nations, creating jobs and wealth for many people, especially young male and female workers.
- Manufacturing jobs are often better paid and more skilled than jobs in agriculture.
- Exporting manufactured goods creates wealth for the country, increasing the GDP.
- Increased wealth might lead to better investment in health, education and infrastructure.
- People working in manufacturing can be exploited, women are often forced to work long hours for low pay in difficult working conditions. Children are often employed, who then miss out on an education.
- Many young people leave rural areas to get jobs in factories in cities, leaving an aging and dependent population behind, which results in rural poverty.

Challenges of an urbanising world

26 Urbanisation trends

1 The increase in the proportion of people living in towns and cities (urban areas).
2 B 48 per cent
3 B 46 per cent
4 In North America and Europe (developed world) the rate of urbanisation is projected to slow down, whereas in Africa and Asia (developing world) the projected rate of urban growth is expected to increase.

27 Megacities

1 C A city with a population of more than 10 million
2 Most of the megacities shown are in Asia, 12 of the 19. Most are found in coastal locations, many are found within the tropics. There are none in Europe.
3 Where the largest city in a country controls disproportionate amounts of economic and political power.
4 Choose one from the following: transport problems as a result of high numbers of people living and working in the city, so public transport can get overcrowded; high volumes of trade and travel cause congestion on the roads, causing longer journey times; high levels of rural to urban migration mean there is often a shortage of affordable housing, which can cause the development of squatter settlements in some developing countries.

28 Urbanisation processes

1 (i) D 2010 (ii) There is an increasing percentage of people living in urban areas. It has increased from 20 per cent in 1980 to over 50 per cent in 2015 – a 30 per cent increase.
2 Lack of well-paid jobs in agriculture in rural areas causes people, mainly the young, to move to cities to find work in newly opened industries/factories. Better housing, services, education and the 'bright lights' in cities attract many young people to move there from rural areas.
3 Counter-urbanisation. People choose to move to less crowded rural areas for a better quality of life and either work from home or commute into nearby cities. OR Deindustrialisation. The closing of heavy industry has meant a loss of jobs causing people to leave the city to find work elsewhere.

29 Differing urban economies

1 Forms of employment that are not officially recognised, such as people working for themselves as street vendors.
2 November, December, January, February 2008–2009
3 People who work in the informal sector do jobs that are not officially recognised and therefore workers have no legal rights, so if they are sick they do not get paid.
4 In developed countries more people work in the tertiary sector providing services, which tend to be more highly skilled and better paid jobs, whereas in emerging countries there is now an increasing number of people working in secondary employment in factories manufacturing goods, which tend to be less well paid. In developing countries there are a higher number of people working in primary industries, which are often low-skilled and low-paid. Increasing numbers of people in the developed world now work in the quaternary sector, in jobs that are highly skilled and highly paid.

30 Changing cities

1 D Rural–urban fringe
2 (i) The large, rectangular buildings indicate that factories could be in this area. (ii) Any one from farming, residential, transport, leisure.
3 There is more space for large assembly plants, so the land is cheaper than in congested CBDs. It is often close to main transport routes, so there is good access for goods and workers. There may be a more pleasant environment that attracts workers.

31 Location and structure

1 Mumbai is a megacity in the Deccan region of India. Mumbai lies on the western coast of India by the Arabian Sea. Mumbai is made up of a group of seven islands and is referred to as the Island City. Your answer will depend on the city chosen.
2 Your answer will depend on the city chosen. For example, Mumbai's coastal location has been very important to its economic development. The deepwater harbour on the Arabian Sea has enabled India's largest container port to develop, which today deals with 40 per cent of India's ocean trade.
3 C Large shops and commercial offices
4 Your answer could refer to: older historic buildings, commercial buildings, department stores, banks, headquarters of large businesses, zone of transition next to CBD – here there are likely to be industries and low-quality squatter settlements.

32 Megacity growth

1 The population increased, the built-up area increased and the amount of rural area decreased.
2 Rural to urban migration. Young people have moved from rural areas, where there were few opportunities in farming, for a good education or to find better-paid work in industries (in Mumbai this could be in chemical industries or call centres). Young people are likely to have more children, which adds to the amount that the city grows.
3 As the population of a megacity increases, as more and more people move into the city, more houses and services will be built. *Answer will depend on chosen city, but likely changes would be:* urban sprawl will occur, i.e. outward expansion on the edges as new suburbs are added, high-rise buildings will be created close to the CBD.

33 Megacity challenges

1 Your answer will depend on city chosen, but is likely to focus on: the lack of affordable housing; people having to live in squatter settlements with poor services and amenities e.g. water; not enough education and healthcare for all.
2 As more people move into megacities, the demand for jobs is high. Not all migrants may have skills and therefore many have to make a living with informal jobs, which require no qualifications and low skills. Many developing countries do not have welfare benefits so people who cannot find a job have to do what they can to survive.
3 (i) In the left of the picture are slums, with very crowded low-quality living conditions, with little space and few amenities. This contrasts with the living conditions to the right of the picture, where there are high-quality apartments with green spaces and leisure amenities, such as tennis courts and swimming pools.
    (ii) The types of jobs people do could determine the type of

living conditions. If a person has a good education and works in the growing service sector – perhaps in a call centre, or in the Bollywood film industry in Mumbai – they would earn enough to afford high-quality housing, whereas a person with low skills may have to work in the informal sector and may only be able to afford to live in a slum.

34 Megacity living

1 Your answer will depend on city chosen, but could include the fact that the poor live in overcrowded squatter settlements with very basic amenities, and often share water taps, whereas the wealthy may live in high-class housing.

2 Migrants often lack education or skills, therefore do low-paid or informal work, meaning they cannot afford to live in high-class housing. Those with a good education and skills, who work in skilled employment such as service industries like call centres, can afford to live in better housing conditions.

3 One advantage of top-down strategies is that they are large-scale, so <u>the problems of the whole city can be tackled together.</u> OR They are government-led, so they can use political power to make landowners sell their land for new developments.

4 Disadvantages of bottom-up strategies are they are small-scale, so they cannot fix city-wide problems such as transport congestion. Schemes that help one community can be resented by other communities.

5 Your answer will depend on city chosen, but may include government schemes and self-help schemes. These involve charities or governments giving grants and materials to people and providing training on how to improve their own living conditions. People may be trained and given the materials such as bricks and cement to construct their own, better-quality homes.

35 Megacity management

1 Your Your answer will depend on the schemes you outline.
- NGOs provide funding for small-scale projects, often aimed at improving living conditions, healthcare and education in local communities.
- One advantage of this is that local people can have a say and be involved in the project.
- These projects are often completed quickly because they are small-scale.
- They can have multiple advantages: improving healthcare means children can attend school, gain more skills and get better jobs.
- They are sustainable because they involve local people, can improve the local economy and have a low environmental impact.
- Disadvantages of these types of scheme are related to their small scale. They cannot fix city-wide problems, and only selected communities benefit, which can cause resentment.
- Improving slums may not always be the right answer. It may be that the government wants the slums cleared to make way for better-quality housing.
- The alternative would be a large-scale, top-down, government-led scheme that addresses whole-city problems, such as providing public transport.

Extended writing questions

36 Paper 1

1 Here are some points you could include in your answer.
- Governments entered into free trade agreements with the World Trade Organization, so that their countries can export more goods and gain wealth and increased GDP from this.

- Previously government-run industries that have been privatised may now be run more efficiently and become more profitable.
- TNCs allowed to set up operations in the country may create more skilled jobs and transfer new technologies to the country. Infrastructure, such as roads, ports and airports, may be improved as a result of TNC investment.
- The lack of, or poor, government policy on working conditions may mean that workers, especially women and children, are exploited in sweatshops or in the informal economy.
- If a government has prioritised improving access to education, many children will benefit from having more skills and be able to find work in better paid, skilled jobs, such as call centres in Mumbai.

The UK's evolving physical landscape

37 Uplands and lowlands

1 (i) C A lowland area made up of mainly sedimentary rocks (ii) Many of the UK's upland mountains are ancient volcanoes, formed from <u>magma at a plate boundary millions of years ago, making igneous rocks</u>.

2 In the uplands you can find glacial troughs (U-shaped valleys), corries and arêtes, which were formed by glacial erosion processes of plucking and abrasion.

38 Main UK rock types

1 A Granite

2 Sedimentary rocks contain fossils. They are made up of layers of deposited sediment, less resistant to weathering and erosion.

3 Slate is formed from clay. Layers of clay are compressed under <u>great heat and pressure to form very strong and very resistant slate</u>.

4 Granite is an igneous rock, formed from magma from past volcanic eruptions. It has been chemically weathered into blocks.

39 Physical processes

1 B A corrie formed by glacial erosion

2 Since the glaciers melted, the rocks in this area will have been weathered by <u>freeze–thaw weathering. In the winter, temperatures fall, causing water in cracks to freeze and expand. Alternating freezing and thawing loosens and breaks the rocks to create scree slopes (broken rock debris) on the sides of the mountain.</u>

3 One from: the river has eroded the valley vertically; mass movement processes on the slopes, such as soil creep and slumping, have made them less steep.

40 Human activity

1 (i) Land has been used for agriculture – fields. Settlement village. (ii) This is an upland mountainous area. The slopes are <u>steep, soils are thin and rocky and the climate cooler. It would be difficult to grow crops here</u>.

2 This landscape was chosen for a settlement because it has a harbour at the mouth of a river, to transport goods and to access fishing resources.

Coastal change and conflict

41 Geology of coasts

1 A concordant coast is where alternate bands of hard and soft rock run parallel to the coast.

2 A discordant coast is where alternate bands of hard and soft rock run at right angles to the coast, whereas <u>in a concordant coast the bands of rock run parallel to the coastline. Discordant coasts have bays and headlands, with features such as caves, arches, stacks or stumps. Concordant coasts tend to have fewer bays and headlands.</u>

3. Choose any two from the following: soft rock cliffs are easily eroded by the sea; cliffs will be less rugged and less steep than hard rock coasts; soft rock landscapes are prone to slumping.
4. (i) 1.53 m (ii) At north Barmston there is a caravan park and a rock groyne has been built on the beach to protect this, therefore slowing down erosion, but there is no coastal protection to the south of the caravan park so erosion would be greater here.

42 Landscapes of erosion

1. (i)

Landform	Stump	Wave-cut platform	Cliff	Stack	Arch	Wave-cut notch
Letter	U	X	Z	V	W	Y

(ii) Your answer should include the appropriate section of the following, for one landform. The headland made of hard rocks would have contained weaknesses, such as cracks. The waves would have crashed into the cracks in the rocks and the force of the compressed air would have enlarged the cracks by hydraulic action, forming caves. Further erosion of the cave by abrasion would have caused the caves in the headland to be breached to form an arch. The roof of the arch would have been worn down by rainwater and the wind (sub-aerial weathering processes). When the arch collapsed a stack would be left behind, which is further weathered and eroded to form a single rock stump.

2. A wave-cut platform is formed at the point where the waves attack the cliff, above which there is an unsupported overhang, which collapses, causing the cliff to retreat and new land at the back of the beach to be exposed as the wave-cut platform.

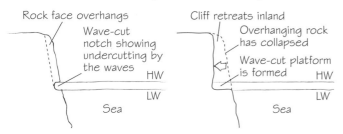

43 Waves and climate

1. **B** Abrasion
2. Destructive waves have a tall wave height and short wave length and they will arrive at the coast in quick succession, whereas constructive waves have a low wave height and long wave length, and will break onto a beach in slow succession.
3. Waves strike the base of the cliffs. The hydraulic action of the water wears the rock in the cliff face away bit by bit through the process of erosion. This repeated action of waves striking the base of the cliff creates a wave-cut notch at the point where <u>the waves attack the cliff, above which there is an un-supported overhang, which collapses, causing the cliff to retreat.</u>
4. Wave height and type will be determined by the fetch – the distance of open water the water has travelled over. The larger the fetch, the taller the waves and the more likely they will be destructive waves. The strength of the wind is another factor. Stronger winds cause taller, more destructive waves.
5. Constructive waves

44 Sub-aerial processes

1. (i) **B** A landslide (ii) Heavy rainfall will have caused the cliffs to become saturated <u>and heavier. The added weight would have caused the cliff to collapse.</u>
2. Sub-aerial processes are land-based processes that alter the shape of the coastline. They are a combination of weathering and mass movement.
3. Cliffs are shaped through a combination of erosion and weathering – the breakdown of rocks caused by weather conditions. Soft rock, such as sand and soft clay, erodes easily

to create gently sloping cliffs. Hard rock, such as chalk, is more resistant and erodes slowly to create generally steep cliffs. Over time, the cliff collapses, causing the cliff line to retreat.

45 Transportation and deposition

1. Your completed longshore drift diagram should look something like this:

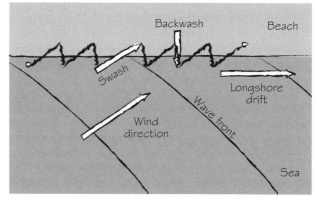

2. (i) Beach (ii) This landform is made up of eroded material (sand, shingle and stones) that has been transported from elsewhere and deposited by the sea. <u>Where there is a drop in coastal energy, often caused by a change in the direction of the coasts (or shelter from a headland), this material is deposited and it forms a beach. Beach material is built up by swash. Longshore drift – a zig-zag movement along the beach caused by the waves approaching the coast at an angle – can 'move' the beach along the coastline.</u>

46 Landscapes of deposition

1. A = bar
2. (i) Spit (ii) Your completed diagram should look something like this:

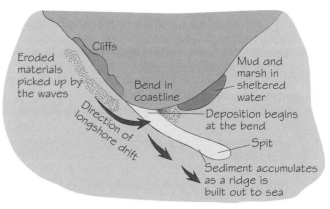

47 Human impact on coasts

1. (i) **A** Retailing (ii) Heavy industry and agriculture bring jobs and wealth to the area.
2. Settlements have been built close to the coast. The weight of the buildings has <u>increased the cliff vulnerability and changed the drainage, which might mean the cliffs get saturated more quickly and create more mass movements.</u> Humans have constructed sea defences, such as groynes, along some coasts. This has helped to <u>protect the cliffs from erosion and cliff retreat, but can increase erosion further along the coastline.</u>

48 Holderness coast

1. Approx 0.565 km in 148 years = 565 m/148 yr = 3.8 metres per year. *Other answers, based on student measurements, should be accepted if working correct.*
2. Softer rocks, such as muds and clays, are more susceptible to erosion; lack of coastal management techniques, such as hard engineering, to protect cliff line from wave attack; large fetch so waves more powerful therefore more erosion.
3. Erosion, mass movement, longshore drift and deposition. In the case of the Holderness coast, the soft boulder cliffs

have been eroded by wave action and sub-aerial weathering processes. Where there has been a lack of sea defences, this has resulted in cliff retreat.

4   Hard engineering has slowed down cliff retreat, and has caused <u>increased rates of deposition around groynes. This has resulted in bigger beaches in some areas.</u>

49 Coastal flooding

1   Choose one of: tourist facility, residential, agricultural, transport

2   Sea levels are rising because of climate change. Temperatures have risen, causing <u>the seas and oceans to warm and expand, leading to a rise in sea levels. Rising temperatures have caused increased melting of glaciers over the land, which has led to more water draining into the seas, raising sea levels.</u>

3   Here are some points your answer could include.
   •   Sea-level rise could accelerate erosion at the coast, leaving homes and businesses close to the coast vulnerable as cliffs retreat.
   •   Property prices may decrease as people may not want to buy houses close to the coast.
   •   It may be difficult to get insurance, or insurance premiums could be high.
   •   Increase in storm events and rising sea levels may cause more coastal flooding.
   •   Beaches, which local people may use for recreation, may be eroded.
   •   Tourism may decrease if the beaches are eroded, affecting tourist-related businesses, jobs and the local economy.
   •   Rising sea levels may mean that the local council needs to invest more in sea defences, which may cause council tax to increase or other local services to suffer.

50 Coastal management

1   Groynes can help reduce rates of coastal erosion by trapping sediment that is being moved along the coast by longshore drift. The trapped sediment <u>builds up a beach, which absorbs wave energy and protects the cliffs and land behind.</u>

2   Rip-rap works by absorbing and deflecting the energy of waves before they reach the defended structure. The size and mass of the rip-rap material absorbs the impact energy of waves, while the gaps between the rocks trap and slow the flow of water, lessening its ability to erode soil or structures on the coast.

3   Beach nourishment has to be repeated fairly often – once a year in some cases – which makes it more expensive over the long term. It also increases deposition further down the coast.

4   Hard defences are normally only used to 'hold the line' i.e. to protect important infrastructure, houses and so on. Examples include the Bacton gas terminal in North Norfolk.

5   ICZM treats the whole coastal area as one area for planned development. It sets out which areas will be protected from erosion by the sea and in which areas the 'do nothing' approach will be used.

Fieldwork: coasts

51 Investigating coasts: developing enquiry questions

1   Swanage is a coastal settlement with a beach and headland, where coastal processes such as erosion, longshore drift and deposition could be investigated. The map clearly shows this is an area where groynes are being used as a coastal defence. This would make a good place to study the impact of coastal management.

2   Possible enquiry questions: How have the groynes on Swanage beach affected the beach? How do the beach deposits vary on Swanage beach?

3   The profile of the beach gradient could be measured from <u>the shoreline to the base of the cliffs. This could be done using a clinometer to measure the gradient along the break of slope along a transect of the beach.</u>

52 Investigating coasts: techniques and methods

1   Field sketches, interviews with local people.

2   (i) It is appropriate to use a clinometer because the beach has uneven variations in gradient. <u>The clinometer would allow readings to be taken at different size intervals as the points of gradient change.</u> (ii) Operation of the equipment is open to human error. A clinometer is not very accurate. Systematic sampling means the break of slopes could be missed and therefore you would not get a true profile of the beach.

3   The geology map shows the contrasting rock types in the area. This could be used to identify areas of soft rock and hard rock to select as sites for primary data collection. It would be a good idea to select sites in contrasting rock types to find out if there are any differences to beach and cliff morphology and to assess the impact of any beach management in the contrasting areas.

53 Investigating coasts: working with data

1   A transect diagram showing profile

2

Concrete sea wall built at the back of the beach to deflect wave energy

Narrow beach where there are no sea defences

Promenade on top of the sea wall for tourists to enjoy the coast

Wider and deeper beach in front of the groyne

Wooden groynes built to trap sediment transported by longshore drift

3   GIS could be used to show the location of the sites overlaid onto a Google Earth map. ARGis could be used to show the sediment size along the beach transect.

4   Here are some points you could include in your answer.
   •   Depending on how the data have been collected for the beach profile, a transect drawing will most likely have been used to present these data. If the data were collected systematically – for example, every 10 metres – this may not be truly representative of the shape of the beach. Break of slope measure may have been a better method, where gradient measurements are taken every time the slope changes, although this is subjective to the person taken the readings.
   •   You should refer to how many transects and how these were sampled.
   •   You should refer to your conclusions and how accurate you think these are based on the data presented.
   •   You should offer descriptions or even a sketch of data collection methods.
   •   Transects could be used to show beach morphology.
   •   Sediment characteristics could use histograms or bar charts to show the different sizes of sediments.
   •   Annotated photographs may be used to visually show the impact of sea defences.

River processes and pressures

54 River systems

1   A long profile shows changes in the height of the course of a river from its source to its mouth. Many rivers have a concave long profile.

2 **B** The width and depth increase downstream

3 (i) A should be V-shaped and C should be wide U-shaped or open V-shaped.

(ii) (a) B, (b) A, (c) C

4 At point A the river is in its upper course. Here the river channel will be shallow and narrow, the sediment size will be large and angular and discharge relatively small, whereas at point C <u>the channel will be much wider and deeper, discharge high and the most of the sediment will be small silt and mud suspended in the channel.</u>

55 Erosion, transportation and deposition

1 **B** The force of the water on the bed and banks of the river removes material

2 Abrasion is the sandpaper-like action of water rubbing against the banks and bed of the river.

3 *Here are some points you could include in your answer.*
Traction (X): Large boulders are rolled along the river bed when the river has lots of energy, usually high near the river's source or during times of flood.
Saltation (Z): Small boulders are bounced along the river bed
Suspension (W): Fine light materials, like sand and silt are carried along by the river flow. The river requires less energy for this and is more common in the middle and lower courses of a river.
Solution (Y): Material dissolved in the water.

56 Upper course features

1 (i) **C** Interlocking spur (ii) In upland areas, small streams begin to develop and erode the landscape. A stream cuts vertically downwards, into the landscape, cutting a small V-shaped valley. Vertical erosion continues to erode the valley in the upper sections. Some parts of the hills tend to stick out into the river valley, resulting in a staggered formation, 'interlocked' together a bit like the teeth of a zip.

2

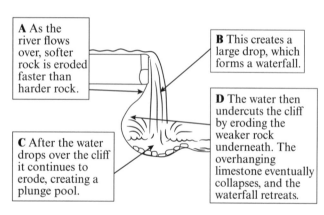

A As the river flows over, softer rock is eroded faster than harder rock.

B This creates a large drop, which forms a waterfall.

C After the water drops over the cliff it continues to erode, creating a plunge pool.

D The water then undercuts the cliff by eroding the weaker rock underneath. The overhanging limestone eventually collapses, and the waterfall retreats.

57 Lower course features 1

1 The main process at A is deposition.

2 The deeper parts of the channel have the highest velocities. The greatest velocity is 0.4 m/sec, which is on the deeper outside bend.

3 A meander is a bend in a river. Meanders form where the gradient of the river is less steep. This means that, as a result of a swinging current, erosion is greatest on the <u>outside bend where the river is deeper. On the inside bend, the river is slow and so deposition occurs.</u>

58 Lower course features 2

1 Levées are formed by the process of deposition. Every time the river floods, sediment and water leave the river channel. The largest amount of sediment is deposited on the banks forming raised embankments called levées.

2 **B** A lowland river flowing through a wide floodplain

3 **C** Meander

4 Deltas form close to the mouth of a river as it approaches the sea or lake. The river loses energy and <u>deposits the sediment</u>

<u>it has been carrying as suspended load in a fan shape. The largest and heaviest sediments are deposited first, closest to the channel, followed by the medium-sized sediments and finally the finest sediments travel furthest into the lake or sea.</u>

59 Processes shaping rivers

1 **C** Slumping

2 The wetter the climate, the higher the river discharge, which will result in <u>more erosion processes and landforms such as waterfalls. This in turn increases the sediment load of a river, resulting in more transportation.</u>

3 The presence of the waterfall in the landscape means that this is an area where there are hard rocks and softer rocks. Soft rock erodes more rapidly forming channels.

60 Storm hydrographs

1 (i) The increase in discharge from the base flow level until it reaches peak discharge (ii) 2 hours (iii) 25 m³/sec (iv) River X is most likely to flood as it has a shorter lag time and higher discharge. This means more water arrives in the river more quickly. The channel is therefore more likely to fill up and overflow.

2 Impermeable rocks will not allow water to percolate into the rocks below the ground. The water then <u>runs off into the river channel more quickly, causing a shorter lag time and a high rising limb.</u>

61 River flooding

1 Here are some points you could include in your answer.
Physical factors.
- There were record high rainfall levels over Cumbria.
- The ground was already saturated so water flowed quickly over the steep Cumbrian mountainsides into the River Derwent and River Cocker.
- The two rivers meet in the town of Cockermouth, therefore increasing the discharge.
Human factors.
- Many old bridges collapsed and blocked the river channels.
- Urbanisation in Cockermouth meant water ran off the impermeable surfaces rapidly into rivers.
- Trees have been removed (deforestation) in parts of the catchment, reducing interception and transpiration and increasing the amount of water in the river catchment.

62 Increasing flood risk

1 **D** River bank erosion

2 Climate change may cause temperatures to increase, leading to evaporation and rainfall; this will cause <u>more river floods. Increased rates of urbanisation – if there is more building and development on floodplains, then more impermeable surfaces means water will run off more quickly into rivers, raising water levels and making flooding more likely.</u> *Could also include deforestation.*

3 River flooding may cause more death and injury to animals and fish. Damage to wildlife habitats may cause a reduction in birds' and animals' food supply. Pollution may damage the land if sewage and chemicals are among the floodwater. Damage to and destruction of plants and trees.

4 More frequent flooding could lead to more homes and property being damaged, resulting in the insurance cost for claims increasing and insurance policies becoming more expensive; increased cost to businesses affected by flooding, for example farmers might lose a crop or factories lose their stock; increased cost to government of building bigger and improved flood defences.

63 Managing flood risk

1 (i) Dam (ii) Example: Dams are built across rivers in the upper course to hold back and store the water. This prevents the water flowing downstream and flooding urban areas. The water can be released gradually after a storm event.

2 Dredging rivers to remove vegetation and rocks to increase the size of the river channel so it can hold more water needs to be done regularly to be effective. OR Disturbs aquatic wildlife habitats. *Other answers possible for other techniques.*

3 Trees slow down the rate at which water arrives in a river by intercepting rainwater on the leaves and branches. This means that the roots of trees have more chance to absorb water, which means that the amount of water going into rivers is reduced, and it goes in at a slower rate (increased lag time).

Fieldwork: rivers

64 Investigating rivers: developing enquiry questions

1 The OS map shows the accessible River Noe and its tributaries in its upper course, where the river should be shallow enough to take measurements. There is the nearby settlement of Edale and a train line, which could allow students to investigate how these could be affected by flooding of the rivers in this area.

2 Example: What are the threats to Edale of river flooding?

3 Discharge on the river could be calculated by measuring the width, depth and velocity of the river channel.

65 Investigating rivers: techniques and methods

1 C A flow meter

2 (i) The tape measure would allow the depth of the river channel to be measured systematically, and an average could then be taken and a cross-section drawn. (ii) The systematic sample means not all of the river bed is measured. It might be that deeper or shallower areas are missed out and therefore it would not be a true representation of the shape of the riverbed.

3 The map shows the areas at risk from flooding, where there are no existing flood defences. You can see that the greatest risk is close to the road to the south of Edale.

66 Investigating rivers: working with data

1 A cross-sectional diagram

2 (i) To calculate the cross-sectional area, you need to multiply the width by the mean depth. (ii) $31.1\,m^2$

3 Google Earth could have been used to show the location of sites. Discharge data could have been entered into a GIS package and overlaid onto the sites.

4 Here are some points you could include in your answer.
- Cross-sectional data would most likely be presented using cross-sectional diagrams. This though only shows the changes along the river bed at the intervals the depth measurements were taken.
- Sediment load could be presented using bar graphs, radar graphs or pie charts. Multiple bar graphs would be needed to show the quantity of the different sediments sizes. Pie charts would only show the proportions and not the actual amounts of the different sediment sizes. Small amounts would be difficult to represent on all these types of graph.
- Discharge and velocity could be presented using bar charts or line graphs if the data are expressed over time.
- Qualitative data might include annotated field sketches. The quality of these will depend on the ability of the person to draw.
- You could include annotated GIS, photos, OS and geology maps.
- Assessment should include how easy it was to interpret the presented data to support conclusions.
- Conclusions reached should be supported with both primary and secondary data.
- Comments would be needed about the suitability and the limitations of how both the primary and secondary data are presented.

The UK's evolving human landscape

67 Urban and rural UK

1 Many young people move out of rural areas to urban areas as there are more job opportunities, higher pay and better services.

2 B The permanent population is declining so fewer people live in these areas to support the services

3 Choose any one of:
- EU Regional Development Fund, provides £2.6 billion to England to help develop poorer regions.
- UK government has designated 24 Enterprise Zones, where tax is cut to attract business. UK government infrastructure projects, such as HS2, are aimed to develop poorer regions in the Midlands and the north of England. *Other answers are possible, e.g. special development zones such as the Northern Power House.*

68 The UK and migration

1 (i) 30 000 (ii) London

2 From 1986 to 1990 the general trend was an increase in population. Between 1990–91 and 1991–92 there was a slight drop in population; from 1992 to 1996 the overall population stayed more or less constant, but migration increased.

3 When people from other countries move to the UK this can have the effect of creating multicultural societies and increasing diversity. This can mean having many different languages and religions, and the range of food types and art, music and sport can increase.

69 Economic changes

1 Globalisation has been an important driver to the changes in much employment in this primary extraction industry, especially the availability of cheaper sources overseas, e.g. Eastern Europe where labour is relatively cheap. It became cheaper to buy coal from overseas where labour is cheaper. There was less demand for coal in the UK, e.g. in power stations, as they began to switch to cleaner alternatives, such as gas. There was little political will to invest in improving performance of coal-fired power stations.

2 (i) D 1990–2010 (ii) There was a large increase in the tertiary sector, particularly banking and finance, which increased by 13.9%. During this time manufacturing decreased significantly by 26.5%. There was a slight decrease in agriculture and fishing. (iii) Industries that provide a service.

70 Globalisation and investment

1 B European Union

2 When foreign companies invest in the UK they create jobs (mainly skilled) for British people. It also increases tax revenue, which can pay for public services.

3 The UK government has encouraged foreign companies to invest in the UK in recent years, in some cases to help fund projects such as new nuclear power stations and property investment. This helps to save the UK government money and helps creates jobs and wealth in the UK.

Dynamic UK cities

71 A UK city in context

1 Your answer will depend on your case study. In the case of Birmingham, the city grew during the Industrial Revolution because of the iron industry and the nearby coalfields, which were an important source of fuel in the iron industries, which later led to the development of a car industry.

2 (i) B Inner city (ii) Traditionally industry would have been located here and low/working class residential.

3 In Birmingham the metal industry polluted the air and the waterways, and the factories became an eyesore.

72 Urban change differences

1. (i) **A** Birmingham (ii) 1% (iii) Example: Birmingham has attracted international migrants because of job opportunities in the growing service industries. Other attractions for migrants include housing and existing multicultural communities. (iii) Causes of migration could be economic: a person may leave one place for another in order to seek a better, well-paid job. OR A person may leave a place for another place for social reasons, for example to be close to family or close to social amenities.
   Could also accept environmental reasons for a better climate, more peaceful, cleaner, safer areas.
2. The most deprived areas are close to the centre (inner city). There are also pockets of high deprivation close to the southern edge of the city. (The least deprived areas are in the northern suburbs.)

73 City challenges and opportunities

1. When the population of an area decreases.
2. **C** Out migration to the suburbs
3. Your answer depends on your chosen city. In the case of Birmingham, the decline of the motor industry in the Longbridge area in the southern part of the city (deindustrialisation) meant fewer job opportunities, so during the 1970s and 1980s this part of the city declined.
4. Gentrification is the improvement of housing in an area that was once run down, by the residents of the area.
5. Urban sprawl is the reason the edges of urban areas have grown, because of the increased demand for living space and larger business, retail and industrial parks. The boom in financial and business services in the CBD has led to gentrification of some inner city areas, where individuals have chosen to live and improve the housing in this area in order to be close to work in the CBD.

74 Improving city life

1. Rebranding is the creation of a new name, term or design for an existing organisation or place.
2. Here are some points you could include in your answer.
   - New and improved homes built and designed to be energy-efficient.
   - Building and developing new housing, workplaces and leisure facilities creates skilled jobs for people. Previously these areas had high levels of unemployment, because of deindustrialisation. Since the Olympics, there are new jobs in retail and leisure.
   - The new facilities can now be used by local people. Building parks has provided green recreational areas for local people.
   - There have been many advantages, in that jobs have been created and the local economy has been boosted. The derelict and polluted post-industrial landscape has been cleaned up and improved, and recreational and leisure facilities have been created for local people.
   - However, many of the advantages have benefited young newcomers to the area. Some of the poorest local people have been forced out of the area, as property prices have increased, displacing the original social problems to other areas.

75 The city and rural areas

1. B Counter-urbanisation
2. (i) Urban residents depend on food produced in nearby rural areas. (ii) People from urban areas use rural areas for recreation or to live in for a more peaceful, safe environment.
3. Urban sprawl can mean rural land is developed and built on and agricultural land and habitats lost. OR Urban visitors in cars can pollute and erode the land and air.

76 Rural challenges and opportunities

1. **C** Farmers try new activities, other than farming, to make money
2. Farmers have seen the value of their agricultural products decrease, because of cheaper imports from abroad, pressure from large supermarkets and changing dietary trends. In order to continue to make money and income, some farmers have diversified into new activities, such as farm shops and farm visits.
3. When people from urban areas buy second homes in rural areas, it puts pressure on housing. This can have the effect of increasing the property prices in rural areas. Housing can get too expensive for local people, especially the young, who are forced out of the area.

Fieldwork: urban

77 Investigating dynamic urban areas: developing enquiry questions

1. (i) Bolton is an urban area with a range of areas of contrasting levels of deprivation, with some areas like Rumworth that have very high levels of deprivation.
   (ii) Do inner city areas have higher levels of deprivation than suburban areas?
2. Example: In order to select the sites for my urban investigation, I first used analysis of secondary data, such as multiple deprivation data and OS maps, to identify contrasting areas/wards. Once the contrasting wards had been identified, then sites within the wards had to be selected. Three streets in each ward were then studied using a systematic sample: in my investigation, every third house/building was surveyed for environmental quality.

78 Investigating dynamic urban areas: techniques and methods

1. **C** Annotated field sketch
2. Scoring can be subjective. What one person finds ugly, the next person might not.
3. The analysis identifies the land use and the image shows the viewer the quality of the environment.
4. The data are useful because they include information about the type of work people do and how skilled they are, and therefore about their pay scales. Professional jobs will be more skilled and more highly paid than elementary and unemployment. The better qualifications a person has, the more skills they will have and the better paid the jobs available to them will be.

79 Investigating dynamic urban areas: working with data

1. Histogram showing multiple bar charts
2. (i) Likely answers include bar graphs, pie charts, stacked bar graphs. (ii) Census data may have to be converted into percentages. If the range of data was too large, exact figures could not be shown.
3. Here are some points you could include in your answer. Radar graph
   - Allows multiple sets of data to be presented on one graph.
   - Allows both positive and negative values to be shown.
   - Trends and patterns stand out clearly, better than in a table of data.
   - If several surveys have been done data can be overlaid. However, this can be confusing and hard to interpret.
   - Often if several surveys have been done, several graphs have to be produced, which be difficult to compare and interpret.
   - An alternative to a radar graph is to construct a modified bar chart with the bipolar score on the x-axis.

Fieldwork: rural

80 Investigating changing rural settlements: developing enquiry questions

1. (i) There are wide variations in deprivation, with some areas such as Penzance that have very high levels of deprivation. (ii) Are the more remote, inaccessible areas more deprived than more accessible rural areas? (iii) Truro and Penzance

2. Example: In order to select the sites for my rural investigation, I first used analysis of secondary data, such as multiple deprivation data and OS maps, to identify contrasting areas/ wards. Once the contrasting wards had been identified, then sites within the wards had to be selected. Three streets in each ward were then studied using a systematic sample: in my investigation, every third house/building was surveyed for environmental quality.

81 Investigating changing rural settlements: techniques and methods

1. C Annotated field sketch

2. Annotations could include: boarded-up buildings that were once businesses – shows jobs have been lost in the area; poor-quality pavements – shows a lack of investment and money from the local council in the area.

3. The data are useful because they include information about the type of work people do and how skilled they are, and therefore about their pay scales. Professional jobs will be more skilled and more highly paid than unskilled jobs and unemployment. The better qualifications a person has, the more skills they will have, the better paid the jobs available to them will be, and the more expensive housing they are likely to live in.

82 Investigating changing rural settlements: working with data

1. A stacked bar graph, clustered bar graphs or pie charts could be used.

2. (i) Likely answers include bar graphs, pie charts, stacked bar graphs. (ii) Census data may have to be converted into percentages. If the range of data was too large, exact figures could not be shown.

3. Here are some points you could include in your answer.
   Radar graph
   - Allows multiple sets of data to be presented on one graph.
   - Allows both positive and negative values to be shown.
   - Trends and patterns stand out clearly, better than in a table of data.
   - If several surveys have been done data can be overlaid. However, this can be confusing and hard to interpret.
   - Often if several surveys have been done, several graphs have to be produced, which be difficult to compare and interpret.
   - An alternative to a radar graph is to construct a modified bar chart with the bipolar score on the x-axis.

Extended writing questions

83 Paper 2

1. Here are some points you could include in your answer.
   - The student only took three environmental surveys in each ward.
   - The student has not made clear how this was sampled and therefore whether the sites surveyed were representative of the wards.
   - Environmental surveys are highly subjective to people's own perceptions and experiences.
   - The range of secondary data was limited to housing type, qualifications and job type. Other types of data could have been used, such as crime levels, and services and amenities.
   - More data would be needed to support the findings.

People and the biosphere

84 Distribution of major biomes

1. A grouping of plants and animals that interact with each other and their local environment; a global ecosystem.

2. C Taiga

3. Tropical rainforests are found in South and Central America, Africa and parts of SE Asia. Their distribution is closely tied to within a few degrees north and south of the equator.

4. Taiga

5. The climate shown in Figure 2 has a large annual temperature range of 42 °C, with winter temperatures falling to −27 °C and summer temperatures reaching 15 °C. This contrasts with a tropical temperature range of only 1.5 °C, where temperatures only get as low as 26 °C and peak at 27.5 °C. The taiga has low annual rainfall compared with high annual rainfall in tropical areas.

85 Local factors

1. B Savanna grassland

2. In the foreground, area X is a savanna grassland with grasses and scrub, whereas in the background, area Y is a snowcapped mountain, with a more tundra environment. It has very little vegetation, just exposed rock.

3. The altitude at area Y is much higher. This would mean temperatures would be lower. The presence of snow would indicate temperatures below 0 °C, which results in very little plant growth and vegetation. The steep mountains slopes would mean thin, rocky, infertile soils, which would be unable to support vegetation.

4. The soils in area X are likely to be more fertile and deeper, because the land is flatter and is at a lower altitude, and the climate would be hotter. This would allow dead organic matter to decompose more quickly into soils.

86 Biosphere resources

1. Choose any from: mineral extraction; energy production; water resources; timber resources.

2. A dam has been built across the river to produce HEP. Could also provide a reservoir for water storage, to irrigate farmland.

3. Tropical areas have high rainfall, which would provide large volumes of water to produce HEP.

4. Tropical rainforests, for example, are important in providing a range of different timber goods (especially hardwoods) that people use for building houses and making furniture. Tropical rainforests can also be a source of fuelwood for local people. Tropical rainforests also act as a biodiversity store of medicines and an important genetic resource (gene pool). The biosphere also directly provides food for people (e.g. fruits and berries) and sometimes water (e.g. baobab tree). It also provides grazing and fodder to animals that humans depend on for food, drink and other products such as clothing.

87 Biosphere services

1. Choose any two of the following: regulating the composition of gases in the atmosphere; maintaining soil health; regulating water in the hydrological cycle.

2. Trees and plants provide leaf litter to the soil, which decomposes and puts nutrients back into the soil.

3. Trees and plants absorb CO_2 from the atmosphere. They therefore reduce CO_2 levels in the atmosphere and reduce global warming. They also give off oxygen as part of photosynthesis. OR The biosphere supports a large biodiversity of plants and animals, which could be genetically important for the discovery of cures for disease.

4. Removing trees and vegetation reduces the amount of water that is returned into the atmosphere through transpiration. This can cause the climate in that area to become drier. Trees and vegetation intercept precipitation. Removing the vegetation means the soils can be exposed to heavy rainfall,

which can erode soils. A loss of soil means a reduction in throughflow and eventually the climate may become drier as a result.

5   Trees and vegetation during photosynthesis absorb CO_2 from the atmosphere and emit O_2.

88 Pressure on resources

1   **A** Food produced is equal to food required to feed the population

2   Over time, both food required and food produced increase arithmetically. At point X, food required increases beyond that of food produced, i.e. there will be a food deficit. Note the difference in the shape of the curves – linear versus exponential.

3   Choose any two from: increasing population; industrialisation; increasing affluence.

4   Boserup's theory is basically that population growth has a positive impact on resource supply. As resources start to run out, people innovate to cope with the problem. When population increases so does food supply, thereby keeping pace with it.

Forests under threat

89 Tropical rainforest biome

1   **C** The soil

2   (i) On the left of the picture there are leaves with drip tips. (ii) In the centre of the picture there are buttress roots of a tree. *Could also mention, the forest has several layers at different heights.*

3   Monkeys have evolved gripping hands and feet so they can balance high up in the canopy to forage for fruit, nuts and berries.

4   The tropical climate has high temperatures over 23 °C and high levels of rainfall all year long. This means that plants can grow all year long. The rainforest is evergreen; there are no seasons. This means there is plenty of food for animals throughout the year, so animals do not have to hibernate or migrate. The tropical climate means there are high levels of biodiversity in tropical rainforest.

90 Taiga forest biome

1   Biotic characteristics are all the living parts of an ecosystem: for example, the plants and animals.

2   The bears have built up fat layers in the summer so they can hibernate in the harsh winters. Their thick fur insulates them against cold temperatures. They migrate to less extreme environments during the harsh winters.

3   (i) The trees have needle-shaped leaves, which remain on the trees throughout the year (evergreen coniferous). This means they can photosynthesise throughout the year and intercept water all year long. (ii) Cone-shaped conifer trees shed snow in the winter so branches don't snap; trees grow close together to reduce damage from strong winds.

91 Productivity and biodiversity

1   **A** The variety of different living species in an area

2   (i)

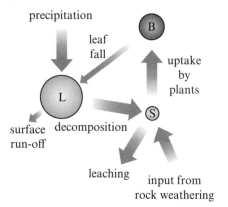

92 Tropical rainforest deforestation

(ii) The tropical rainforest has very high nutrient storage in the biomass compared with smaller biomass storage in the taiga. The tropical rainforest and the taiga both have low soil nutrient storage. The taiga has high nutrient storage in the litter, whereas the tropical rainforest has very low litter storage.

3   The leaves and other dead organic matter that drop onto the forest floor in the tropical rainforest are rapidly decomposed because of the hot and wet climate, which makes conditions for decomposition ideal.

92 Tropical rainforest deforestation

1   In 1975 the area shown was almost entirely covered in forest/vegetation, apart from two roads in the north and east of the area. By 1985, rows of forest (about 30%) had disappeared adjacent to the roads in the eastern part of the area, whereas by 2001 large areas (almost 75%) of the forest had been removed, extending south and west of the area.

2   (i) The area may have experienced deforestation. Hardwood trees may have been removed for use as timber. (ii) The area could have been cleared to make way for plantations, where tropical crops, such as bananas, soya or biofuels, are grown in large quantities for commercial purposes.

3   Climate change could make the tropical rainforests drier and even lead to desertification. This could mean the specialised plants and animals that live in the rainforest may not be able to survive and may become extinct. Drier and hotter conditions may make forest fires more likely, which may destroy the forest and animal habitats.

93 Threats to the taiga

1   **C** Acid rain

2   Removing the trees for softwood removes the habitat and food supplies for animals. This can cause the animals to become extinct. Removing trees leaves the soil unprotected. Soils can blow away without vegetation cover, causing them to become thin and degraded. This can prevent any plant growth.

3   Here are some points you could include in your answer.
   - Future climate change could see temperatures increase in the taiga regions. This may cause the specialised adapted animals to become endangered or extinct.
   - Warmer winter temperatures will allow new diseases and pests to spread to the taiga. Taiga plants and animals will not have resistance to these, so species will die out.
   - There will also be increased frequency of forest fires, which taiga vegetation is not adapted to.
   - Other threats are from commercial development of the taiga from logging, and exploitation of minerals and fossil fuels.

94 Protecting the rainforest

1   From 1988 to 1995 deforestation in the Brazilian Amazon fluctuated from just over 20 000 km² in 1985 to just under 15 000 km² in 1994. Deforestation peaked in 1995 at 28 000 km² after which it decreased until 2002 when it rose to 25 000 km². Since 2002 deforestation has steadily decreased to a low level of 8 000 km² in 2009.

2   Deforestation rates may decrease if the government in that country introduces policies to protect the rainforest. An example of this is in Brazil, when the government withdrew grants to poor settlers to clear rainforest land.

3   (i) CITES or REDD *(ii) CITES has huge global influence, with 181 countries signed up to it. This puts pressure on countries not to hunt for endangered species such as the rhino and therefore helps protect this species. However, it is very difficult to check countries are enforcing the CITES rules and 1000 rhinos were still killed for their horn in 2014 by poachers in South Africa. REDD (reducing emissions from deforestation and degradation) involves offering incentives for changing the way forest resources are used. It offers a way of reducing CO_2 emissions through paying for actions that prevent forest loss or degradation. Examples include carbon trading and paying

for forest management. This helps support forest and habitat conservation, relatively cheaply. There are concerns though about the effects negative impact REDD payments might have on local communities, who may not then be able to develop the local resources and limit agriculture in the area.

95 Sustainable tropical rainforest management

1 **C** The Ecuador proposal is one example of how rainforest can be conserved for the future
2 (i) Ecotourism or sustainable farming (ii) The challenge is making sure the environment is protected, but not at the expense of the economic development of the area and the quality of life for the people living there.
3 Ecotourism allows small numbers of people to visit the rainforest and learn about it and why it needs protecting. Visitors will bring money and jobs for local people and the economy but won't damage the environment.

96 Protecting the taiga

1 (i) National parks prevent development and therefore stop deforestation and protect the area. People are encouraged to enjoy but respect the area. (ii) Animals native to the taiga migrate long distances, often out of the national park, where they will no longer be protected. OR The taiga ecosystem is easily damaged by pollution, especially acid rain, which may be created outside of the national park, but is blown in by the wind.
2 Taiga forests are very fragile ecosystems and, if damaged, take a long time to recover. The plants take a long time to grow back because of the lack of nutrients and cold winters. There is a lot of pressure on the taiga forests as they are a valuable softwood resource and in some areas large stocks of oil and gas have been discovered and there is pressure to clear the forests to extract this oil and gas.
3 Some people believe that the taiga forests should be protected because they are very fragile ecosystems, with many endangered species that have adapted to the environment. Others believe that the taiga forests contain important resources, which should be extracted. For example, in Siberia and Alaska oil and gas have been found under the taiga forests. The development of these resources will help economic development of these countries and provide energy security for the regions.

Consuming energy resources

97 Energy impacts

1 Coal, oil, natural gas
2 **C** Nuclear power
3 Some people believe that wind turbines look ugly, are noisy and spoil the look of the landscape. They take up rural land and can interfere with bird migration routes.
4 Removing the vegetation on the surface of the ground and clearing large areas of land to develop the mine. This can cause the landscape to be scarred and it disturbs wildlife habitats.

98 Access to energy

1 **B** 15%
2 All of these renewables depend on the natural resources available. Few places have access to geothermal energy (e.g. Iceland); similarly, not all places get many hours of sunshine to use solar power, enough wind or are suitable for growing biofuels. More countries have access to rivers, lakes, etc. to make use of hydroelectric power, hence it is the largest renewable energy source.
3 (i) Nigeria consumes fewer oil resources than the USA. Nigeria consumes <1 million barrels of oil per day compared with >10 million. The USA is one of the highest consumers of oil, whereas Nigeria is one of the lowest. (ii) Nigeria is a developing country with lower levels of national wealth. Despite having oil resources, there is low demand for oil, as there are fewer people with cars than in the wealthy USA.

99 Global demand for oil

1 The world oil demand increased from 1990 to 2025 by 40 million barrels per day. The USA remains the highest consumer of oil, but consumption has reduced since 2005 and is expected to continue to decrease. China has had the biggest increase in oil consumption.
2 South East Asia and China have increased their demand for oil because of the increased rate of industrialisation in these countries, which uses large amounts of oil. China also has a large population of over 1 billion people, who have more wealth as a result of industrialisation and are therefore consuming more goods, especially cars, which use more oil.
3 If there is continued conflict in the Middle East then this could hinder the trade of oil from this area, causing the price to increase. If relations between Middle Eastern countries and industrialised countries improve, this could increase the trade between these countries and bring down the price of oil. Also the stocks of oil are likely to decrease over the next 50 years. This could result in the price of these resources increasing if the demand remains high.

100 New developments

1 **A** Tar sand
2 Extracting oil from tar sands or fracking for shale gas will increase the supply of these energy sources and this could keep the prices down for the short term.
3 Here are some points you could include in your answer.
   Advantages
   - Drilling for oil in the Arctic will mean an increase in supply and therefore it will keep the cost of oil down.
   - Large oil companies and workers will profit from sales of oil from the Arctic.
   - Industrialised countries such as the USA and China can continue to use oil as a source of energy and fuel.
   Disadvantages
   - Extracting oil can cause pollution, for example the BP oil spill in the Gulf of Mexico. There are many endangered plant and animal species that could be affected by pollution in this area.
   - It is disputed who owns the resources in this region and this could cause conflict between nations.
   - Because of the very hostile conditions, extracting oil in the Arctic could be very expensive, causing the cost of oil to increase.
   Alternatives
   - Move away from problematic fossil fuels to cleaner, more sustainable options such as wind power, or HEP.

101 Energy efficiency and conservation

1 Thinking carefully about how energy resources are used now to ensure that they have a sustainable future; reducing demand to save supplies.
2 (i) New homes can be built with thick cavity walls and filled with mineral wool insulation to prevent heat loss. (ii) *choose one of*: Replacing existing windows with double or triple argon-filled glazing to prevent heat loss, adding loft and wall insulation to existing homes, lining hot water boilers with heat retaining jackets, switching to condensing boilers, switching off appliances, installing solar panels, and using energy-saving light bulbs and smart meters.
3 (i) The bus runs on hydrogen, which does not produce any CO_2. Switching from diesel or petrol buses to hydrogen will therefore reduce carbon emissions. (ii) In order for people to use hydrogen-fuelled vehicles, there will need to be a large network of hydrogen filling stations, which would be very expensive to build. Also, hydrogen cars are very expensive to buy and not many people can afford them.

102 Alternative energy sources

1 The USA was heavily dependent on fossil fuels in 2012: 87 per cent of their energy mix was made up of fossil fuels. The problem with using fossils fuels is that <u>fossil fuels contribute to a country's carbon footprint as they emit high amounts of CO_2, which contributes to global warming. Also fossil fuels are finite and current stocks will soon run out.</u>

2 The USA could expand its use of renewable energy sources, such as HEP.

3 (i) HEP (ii) Large dams need to be constructed across rivers and large areas flooded, which can destroy aquatic habitats and displace people living in the area.

103 Attitudes to energy

1 Here are some points you could include in your answer.
   • The scientific community is split, with many scientists in favour of developing cleaner renewable sources of energy, which do not increase CO_2 emissions and contribute to climate change. Not all scientists agree, however.
   • TNCs – many oil and gas companies are not in favour of developing renewable energy further, as this may decrease the demand for oil and gas and cut the profits of these companies. Many oil and gas companies are now, however, diversifying into renewable energy.
   • Many governments have signed protocols to cut their carbon emissions and therefore are devising policies to encourage low-carbon renewable energy sources.
   • The public – as people learn about the environmental problems of oil and gas, many are keen to use cleaner renewable energy sources. As people become wealthier they can choose cleaner alternatives.

Extended writing questions

104 Paper 3 (i)

1 Here are some points you could include in your answer.
   • Nationally, large HEP projects can produce large amounts of clean, green power for industry, which can help further develop a country economically and tackle some of the problems in the overpopulated cities.
   • Brazil has abundant HEP resources in the Amazon basin. This could reduce its need to import expensive oil and gas and enable them to be more self-sufficient in energy.
   • Jobs will be created locally in dam construction and in the power stations. The area's local economy will be boosted.
   • However, local ancient tribal traditions may be lost with the influx of migrant workers. The local environment will change and resources such as fish stocks will be depleted.
   • Brazil may feel the impact of climate change if more of the rainforest is lost through industrial development.
   • 78.8 per cent of electricity in Brazil comes from HEP.
   • Fish species such as the black and white-patterned zebra pleco fish are endangered.
   • 40 000 local people will be affected by the construction of the dam.

105 Paper 3 (ii)

1 Your answer must consider all three alternative options and establish a clear argument about the meaning of the 'best long-term' plan. There is no preferred option. All options can be justified. The balance of the case made will vary according to the chosen option.

Option 1: Build the Belo Monte dam and further large HEP schemes to produce more HEP

This can be justified by suggesting that Brazil's long-term economic development will depend on the development of energy resources. HEP is a clean, green energy source that will not contribute to global warming. It is also an infinite energy resource that will not run out. Development of the dam will create skilled jobs for local people and migrant workers, which will boost the local economy and the quality of life for people in the area. There are environmental concerns with this option, such as the flooding of large areas of rainforest, interrupting the flow of the river and disturbing aquatic ecosystems. With careful planning and management, many of these concerns can be addressed.

Option 2: Make the area a national park and increase the amount of energy produced from biofuels

This can be justified by suggesting that local and global resources from the tropical rainforest will be conserved for the long term. This would mean that the rainforest resources could not be used for commercial benefit and this could hinder economic growth in Brazil. However, the national park could attract tourists, which could help boost the local economy. Developing biofuel is a sustainable option: it is a carbon-neutral fuel; the climate and environment are suited to growing the crops needed; this would require large areas of land, but it could be argued there is sufficient space in Brazil to do this.

Option 3: Develop smaller HEP dams and solar and wind power

This can be justified as a more bottom-up approach, which might mean less local opposition and less damage to the environment compared with the top-down mega dam option. Solar power may not be very efficient in some cloudy parts of the rainforest. Large areas of rainforest would have to be cleared to make way for solar farms. Wind farms may not be suitable deep in the sheltered rainforest regions, which contain dense vegetation. Development may be slower but more sustainable, as solar power, wind power and HEP do not emit any greenhouse gases and will not be finite in supply.

Skills

106 Atlas and map skills

1 (i) Brazil should be shaded as an upper-middle income country, Kenya as a low-income country, Saudi Arabia as a high-income country and Ukraine as a lower-middle income country. (ii) **D** Choropleth

2 The highest income countries can be found in the continents of <u>North America and Europe in the Northern Hemisphere, where income levels are US$11 116 or more. The lowest income countries are mainly found in Africa and Asia.</u>

107 Types of map and scale

1 Most hot deserts occur close to the tropics of Cancer and Capricorn. The greatest extent of hot desert is in a belt across North Africa and the Middle East. There are no hot deserts in the higher latitudes of the Northern Hemisphere, although the map shows hot deserts stretching far to the south in South America. Hot deserts in the Americas are most frequently located along the eastern coastline. A large part of central Australia is hot desert.

2 (i) About 135 m (ii) 300 m

108 Using and interpreting images

1 (i) **B** Oblique aerial (ii) Advantage 1: It is easy to compare different land uses. Advantage 2: <u>Choose any one from: indicates height and density of settlements; outline of transport network is clear; photo available for use quicker than map; provides record of the area that day.</u>

2 (i) The river is flowing northwards into the sea. (ii) The contours and spot heights give the height of the land.

109 Sketch maps and annotations

1 (i)

(ii) A river cliff is the steep-sided bank of the river which forms <u>when fast-flowing water erodes the sides of the bank, causing the top of the bank to collapse into the river</u>. The emphasis of your answer should be on explanatory detail.

2

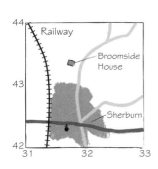

110 Physical and human patterns

1 (i) Alnwick has a <u>nucleated settlement</u> shape because <u>it is clustered around main roads and sandwiched between the A1 and the B6341 which runs from the south-west of Alnwick</u>. You must use evidence from the map.

(ii) You must refer to evidence from the map or give an explanation. Alnwick will have difficulty expanding because there is coniferous woodland to the north-west of the settlement, in GR 1713. This will need to be deforested to allow expansion. There is farmland to the east of the A1, therefore land is valuable for primary industry. The flow of the River Aln is to the north and north-east, so will inhibit development; high relief to the west and south-west, which is difficult to build on.

111 Land use and settlement shapes

1 **B** Spread out

2 (i) 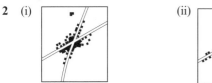 (ii)

3 (i) Sidford is a nucleated settlement clustered around roads.
(ii) The land use is very rural. Brook Farm is located to the …
You should include both physical and human land uses and try to use evidence from the map. Here are some points you could include in your answer.
Physical: Sidford lies in a valley with steep land to the north-east and north-west; very high land (198 m) to the west; main river flows from north, tributary from north-west.
Human: Main settlement is Sidford; transport links into Sidford, A375 to the north; farming settlement – Brook Farm sandwiched between Core Hill and Castle Hill to the north-west of Sidford.

112 Human activity and OS maps

1 School or infirmary
2 Bus, because there is evidence of a bus station
3 Choose any two from: post office; pub; church
4 Shilbottle is surrounded by a rural landscape. The many farms around the area help prove this, for example South East Farm located to the south east of Shilbottle. The map shows <u>a lack of built-up areas – it is mainly an agricultural landscape. Transport routes are limited to roads less than 4 m wide, which supports limited rural population. Contour lines show that this is an upland area and therefore the landscape unable to support the growth of large settlements</u>. You must use evidence from the map.

113 Map symbols and direction

1 (i) Parking, mixed woodland, public house, bus station
(ii)

2 (i) North-east (ii) 224 m (iii) Triangulation pillar on Beacon Hill indicates highest point of 224 m

114 Grid references and distances

1 (i) <u>Deciduous</u> (ii) 113911 (iii) 4 km (iv) B Claypitts Farm
2 C 091923
3 (i) 6 km (ii) Telephone (iii) 112909 (iv) 087919

115 Cross-sections and relief

1 The patterns match as below.

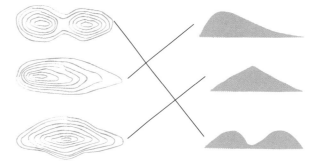

2 The land to the east of the River Otter is <u>steep</u>. It rises to a maximum height of <u>99 m</u> above sea level. In comparison the land to the west of the river is <u>flatter</u> and rises to a height of approximately <u>70 m</u> above sea level.

3 (i)

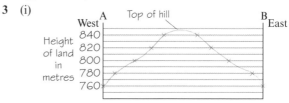

(ii) Try to refer to the data. The western side of the hill rises evenly in increments of 20 m. The top of the hill is relatively flat and is approximately 840 m above sea level. The downward slope of the top of the hill is a mirror image of the left (west) side.

116 Graphical skills 1

1 (i) This should be halfway between the 140 and 160 lines on the vertical axis and directly above the 2011 point on the horizontal axis. (ii) 70 per cent (iii) Eight years (between 2000 and 2008)

117 Graphical skills 2

1　(i)

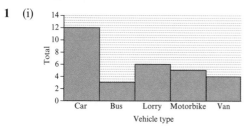

　　(ii) 30 vehicles

2　(i) Population pyramid (ii) The country has a high birth rate: the base of pyramid is very wide, there are lots of children in the 0–4 age band. There is a youthful population – the pyramid narrows considerably above the 30–34 age band. The top of the pyramid is very narrow, indicating a high death rate. Life expectancy in men is lower than in women.

3　The traffic data cannot be shown as a line graph as it does not show a change over time. Line graphs are used to show trends or patterns to see whether there is a correlation between two sets of data.

118 Numbers and statistics 1

1　(i) 17.8 per cent (ii) 6.5 per cent (iii) A line graph would be the best way to represent India's population data because the data show changes over time.

2　$833\,087\,662 \div 377\,105\,760 = \underline{2.2}$. This means for every 1 person living in an urban area in India <u>there are 2.2 people living in rural areas. Therefore the ratio of rural to urban population in rural areas in India in 2011 is 2.2 : 1</u>.

119 Numbers and statistics 2

1　(i) 18.03 m³/s (ii) 18.6 m³/s (iii) The values in order are: 10.7, 14.8, 17.7, 18.6, 22.6, 22.9, 23.2. There are seven numbers, so the median value is the 4th value: 18.6. The lower quartile number is the 2nd value and the upper quartile is the 6th value. The interquartile range is the difference between the lower and upper quartiles. In this case the answer is 8.1.
　　(iv) The interquartile range is a measure of dispersion around the average. It omits the very extreme values. The discharge data could be used <u>by authorities planning flood management schemes.</u>

Notes

Notes

Notes

Notes

Notes

Published by Pearson Education Limited, 80 Strand, London, WC2R 0RL.

www.pearsonschoolsandfecolleges.co.uk

Copies of official specifications for all Pearson qualifications may be found on the website: qualifications.pearson.com

Text and illustrations © Pearson Education Limited 2016
Produced by Cambridge Publishing Management Ltd
Typeset and illustrated by Kamae Design Limited, Oxford
Cover illustration by Miriam Sturdee

The right of Andrea Wood to be identified as author of this work has been asserted by her in accordance with the Copyright, Designs and Patents Act 1988.

First published 2016

22
11

British Library Cataloguing in Publication Data

A catalogue record for this book is available from the British Library

ISBN 978 1 292 13376 8

Printed and bound in Great Britain by Bell and Bain Ltd, Glasgow

Acknowledgements

Content is included from Rob Bircher, David Flint, Anne-Marie Grant, David Holmes and Kirsty Taylor.

The author and publisher would like to thank the following individuals and organisations for permission to reproduce copyright material.

Figures

Figure on page 3 adapted from Global mean sea level estimates, records and projections, 1800–2100, Time series IPCC, FAQ 5.1, Figure 1 from Climate Change 2007: The Physical Science Basis. Working Group I Contribution to the Fourth Assessment Report of the Intergovernmental Panel on Climate Change [Solomon, S., D. Qin, M. Manning, Z. Chen, M. Marquis, K.B. Averyt, M. Tignor and H.L. Miller (eds.)]. Cambridge University Press, Cambridge, United Kingdom and New York, NY, USA; Figure on page 68 adapted from Regional Internal Migration, year ending June 2010, National Statistics, Office for National Statistics licensed under the Open Government Licence v.3.0.; Figure on page 68 adapted from Figure 2 Population change in the UK, 1986–1996, National Statistics, Office for National Statistics licensed under the Open Government Licence v.3.0; Figure on page 69 from Coal Authority, Contains public sector information licensed under the Open Government Licence v3.0; Figure on page 70 adapted from UKTI Inward Investment Report 2014 to 2015 (Online viewing), UK Trade & Investment, Contains public sector information licensed under the Open Government Licence v3.0; Figure on page 72 from Figure 1 Population change in the West Midlands, 2008–2009 Source: Office for National Statistics, Office for National Statistics licensed under the Open Government Licence v.3.0; Figure on page 72 adapted from Figure 3 Deprivation in Birmingham, 2010, Crown Copyright: Produced by Birmingham Public Health Information & Intelligence (Dec 14) © Crown Copyright and database rights 2014 Ordnance Survey 100021326; Figure on page 78 from Census 2001, Key Statistics Table KS19, 'Rooms, amenities, central heating and lowest floor level', Crown Copyright, Office for National Statistics licensed under the Open Government Licence v.3.0.; Figure on page 80 adapted from Areas of deprivation in Cornwall, The English Indices of Deprivation 2015 © Crown Copyright, Office for National Statistics licensed under the Open Government Licence v.3.0.; Figure on page 94 from Graph: Deforestation in the Brazilian Amazon Graph. With permission from Mongabay; Figure on page 98 from U.S. Energy Information Administration (2015).

Maps

Ordnance Survey Maps, pp.46, 51, 58, 64, 71,108, 110, 113 © Crown copyright 2016, OS 100030901 and supplied by courtesy of Maps International created by Lovell Johns Limited; Map on page 65 Contains public sector information licensed under the Open Government Licence v3.0, Environment Agency.; Map on page 113 from Ordnance Survey, OS and Maps International created by Lovell Johns Limited.

Photographs

(Key: b-bottom; c-centre; l-left; r-right; t-top)

Alamy Images: A.P.S. (UK) 30, ACE STOCK LIMITED 53, 128, Andrew Holt 41, ARCTIC IMAGES 101, Cath Harries / Alamy Stock Photo 74, Dave Porter 50, David Burton 49, David Gowans 39b, David J. Green 81, David Pick 45, Eddie Gerald 65, Gavin Haskell 39t, Heritage Image Partnership Ltd 2, Kevin Britland 46, Lana Rastro 78, Mike Greenslade 108, Nature Picture Library 85, Pulsar Images 33, Rob Hawkins 52, Sue Cunningham Photographic 86, Tremorvapix 40b, Universal Images Group North America LLC 92, walespix 67, Zach Holmes 15; **Courtesy Mrs Robina Herrington of Skeals**: 47; **Fotolia. com**: annacurnow 38, Beautifulblossom 89, byrdyak 90, Chrispo 44, Jochen Scheffl 13, Pipo 11, sara_winter 59; **Getty Images**: Alex Dellow 71, Mario Tama / Staff 7; **Reuters**: Ricardo Moraes 120; **Shutterstock.com**: aricvyhmeister 63, Katherine Welles 96, Kevin Eaves 40t, PRILL 20, Stephen Meese 97, Timothy O'Leary 102

All other images © Pearson Education